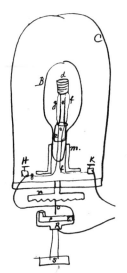

Edison's

Electric Light

W9-BNB-259

Edison's

Electric Light

Biography of an Invention

Robert Friedel & Paul Israel
with Bernard S. Finn

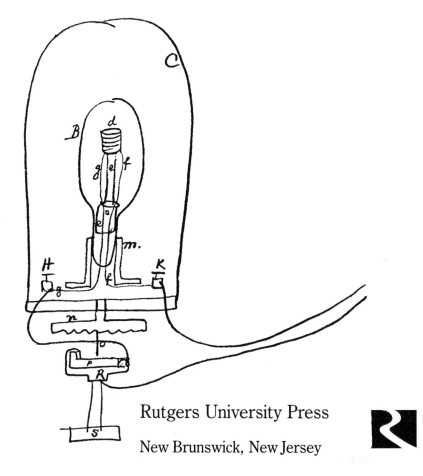

Rutgers University Press

New Brunswick, New Jersey

First paperback edition, 1987
Second printing, cloth, 1987

Library of Congress Cataloging in Publication Data

Friedel, Robert D. (Robert Douglas), 1950–

  Edison's electric light.

  Bibliography: p.

  Includes index.

  1. Electric lamps, Incandescent. 2. Edison, Thomas A. (Thomas Alva),
1847–1931. I. Israel, Paul. II. Finn, Bernard S., 1932– . III. Title.
TK4351.F75   1985   621.3′092′4 [B]   85–2036
ISBN 0-8135-1118-6

    0-8135-1254-9 (pbk.)

Distributed exclusively in Canada by Kuska House, Kingston, Ontario.

CONTENTS

ACKNOWLEDGMENTS

This work began as a study commissioned by the U.S. National Park Service's Edison National Historic Site on the occasion of the hundredth anniversary of Thomas Edison's invention of the incandescent electric light. It owes its inspiration, as well as much material support, to the Site and to the capable and dedicated individuals in whose care lie the treasures of Edison's West Orange, New Jersey, laboratory and home.

Over several years of work, this study has been very much a cooperative endeavor, in which the authors have shared the labors (and the fun) of every aspect of the enterprise. It is fitting, however, to note the fundamental division of labor reflected in this final product. The many months of intensive research in the Edison archives at West Orange was the work of Paul Israel, whose position with the Thomas A. Edison Papers (a project to edit Edison's papers in microfilm and book editions) has allowed him to continue over the years to make the best use of the riches of those archives as this work has moved forward. The text and the interpretive framework in which it is set were the work of Robert Friedel. The impetus for our efforts was provided initially by Bernard Finn, whose enthusiasm, encouragement, and material support, no less than his written contributions, represented by the short essays between various chapters, were essential to our work.

No project of this kind, extending as it has over a period of years, could have proceeded successfully without the generous assistance of many individuals, only a few of whom can be named and thanked here. Most essential, of course, has been the help and generosity of the National Park Service and the staff of the Edison Site, including Ray Kremer, William Binnewies, Roy Weaver, Reed Abel, Leah Burt, Frank McGrane, Anne Jordan, and Edward Pershey. The staff of the Edison Papers Project, both at West Orange and at Rutgers, in New Brunswick, has also been indispensable, and our thanks are extended to friends and colleagues associated with that project, including Susan Schultz, Tom Jeffrey, Leonard Reich, Toby Appel, and Reese Jenkins.

The provision of the illustrations in this work presented some special challenges, both technical and financial. Technical challenges were overcome through the expert assistance of Gilbert Acevedo and, especially,

Joyce Bedi whose generosity with her time and expertise was a contribution for which we are particularly grateful. Financial burdens were relieved by funding from the Office of Research and Sponsor Programs, Rutgers University, and the Department of History, University of Maryland, College Park, which we acknowledge with thanks.

A number of individuals at other institutions have been generous with their help and advice, and if we neglect to name them all, we beg their forbearance. John Bowditch was our friendly and knowledgeable host at the Henry Ford Museum and Greenfield Village, Dearborn, Michigan, where others of the staff of the Museum and the Henry Ford Archives were also ready to lend a hand. Professor P. L. Kirby of the University of Newcastle-upon-Tyne was a most helpful and courteous guide to the documents relating to Joseph Swan's electric light research, and the staff of the Tyne and Wear County Council Archives Department also deserves thanks for making these documents available. Diane Vogt provided thoughtful help in sharing relevant materials from the Archives of the Corning Glass Works. Finally, thanks are extended to several individuals in the Department of History of Science and Technology and the Division of Electricity of the Smithsonian Institution's National Museum of American History for their kind assistance at several phases of this project, with special mention of the help provided by Anastasia Atsiknoudas and Ray Hutt.

Our work has been faithfully supported over the years by the interest, understanding, and liberality of our employers, in particular the Institute of Electrical and Electronics Engineers and the Thomas A. Edison Papers.

Robert Friedel
College Park, Maryland

Paul Israel
New Brunswick, New Jersey

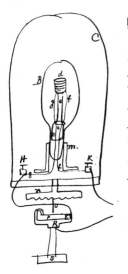

INTRODUCTION

History and

Invention

In the nineteenth century, when "technology" was an intellectual concept and not a popular catchword of politicians and economists, "invention" was the widely accepted mainstay of material and industrial progress. For most of the century, the inventor was a figure worthy of both popular esteem and commercial respect, and the fruits of his labors were viewed as the symbols and sources of the new industrial culture's special strengths. In invention was seen that melding of brain and hand which betokened a new, more democratic and egalitarian world, in which material progress was simply the most obvious manifestation of a spiritual advancement that would eventually extend to all the races of the earth. At least, such was the American creed that propelled the unchecked expansion of industrialism.

The workings of invention normally presented no mystery to an informed citizen of the last century. "Ingenuity," and not "genius," was invention's parent, and ingenuity was within the grasp of any man who had his eyes open, his mind alert, and his hands ready. The same talents at work in the Yankee's marketplace sharpness or meetingplace glibness could be applied in the workplace and machine shop as well. To the ordinary American, the models that stood row on row in the Patent Office were more representative of his country's intellect than any books in a library or paintings in a gallery. After all, these models were generally made by common men (and a few women), graced by no special privilege or education, but simply a shade cleverer or a moment quicker or just a bit nimbler than their fellows. Americans shared something of the faith espoused by Samuel Smiles, when he ascribed the accomplishments of the great inventors and engineers of the day to clear thinking, hard work, and common virtue.

There appeared from time to time, however, exceptions—individuals whose inventiveness seemed to transcend the ordinary and place them and their work on a higher plane. Such individuals were put into the pantheon of cultural heroes, their names to be conjured up to evoke the spirit of progress or, at least, the profitableness of creative enterprise. By the time America had reached its hundredth birthday, in 1876, a fair number of inventors could be said to have achieved such heroic stature and their names were familiar to every schoolchild—Franklin, Whitney, Fulton, and Morse were perhaps the most obvious examples. At that time, however, there was about to emerge another of these exceptional inventors, whose capacity for creating not only the useful and the clever but, occasionally, the miraculous as well, would earn him the title of "Wizard." This was, of course, Thomas Alva Edison, and it is in his wizardry, more than any other single thing, that we can see the beginnings of systematic invention that would regularly go beyond the limits of full comprehensibility for the common man. In Edison we find the transition from the common, ingenious invention that seemed to move much of the world forward in the nineteenth century to the specialized, scientific technology that was to be a dominating social and economic force in the twentieth.

It is important to remember that a transitional figure is just that—an individual who is neither consonant with the old order nor fully integrated into the coming one. This is certainly true of Edison. The "Wizard of Menlo Park" was quickly recognized (when he was barely thirty, in fact) as someone who did things differently than the inventors to whom he might be compared. The "invention factory" of Menlo Park was obviously unlike anything in the ken of even the best informed American. And the kind and number of things that seemed to emerge with diurnal regularity from that little New Jersey village simply brooked no comparison. There was certainly something at work here beyond ingenuity and hard work applied in a useful way.

On the other hand, those who would see in the self-taught, unpolished (and occasionally uncouth) former telegrapher, with his white-clapboard two-story laboratory and his "gang" of faithful mechanics and other helpers, a research and development manager in the twentieth-century mold are far from the mark. The workshops and men at Menlo Park did not constitute a technical laboratory of the corporate type, nor did their leader bear any resemblance in style or action to the successful technocrat of a later day. Edison, in fact, showed himself later in life to be constitutionally unable to operate in the style of the new century. There were others of his generation, such as Elihu Thomson or Frank Sprague, who proved to be better able to make the adjustment to the corporate and professional environment of the twentieth-century engineer. Edison's ambiguous position between the old and the modern ways of invention and progress was reflected in his own time by the picture, still familiar more than a half-century past his death, of the folksy, middle-American hero surrounded by the aura of technical genius.

That such an individual should hold a fascination not only for his contemporaries but also for scholars and laymen of a later time is no surprise. Both the life and the myth are too rich and too important not to
have a sustained influence. What is surprising, however, is the relative
neglect of the internal workings of the Edisonian achievement and, in
particular, of the very stuff of his contribution—his inventions. This is
not to say that Edison's inventions have not been much written about, for
they have, but rather that what has been written has too often belonged
more to myth-making than to scholarship. But even when one enters the
realm of Edison scholarship—a not inconsiderable territory—rarely can
one find true probing questions asked about the act of invention itself,
and the questions that are asked tend to be answered with less than reliable evidence.

The reasons for this neglect are complex. One contributing factor is
the nature of historical scholarship, for only recently has the tackling of
such technical issues become an accepted part of the historical enterprise. Another is in the nature of the evidence, for, while it is certainly
available in great quantity, it is of a type that most humanist scholars in-
stinctively shy away from, and it has long been organized in a manner
uninviting to all but the most persistent researcher. But, ultimately, the
primary reason lies in our cultural perceptions of invention. In the nineteenth century, invention called for no explanation, since it was not seen
as an intellectual endeavor. In the twentieth century, attention is directed toward technology, and not invention, hence efforts are made to
explain the institutional, economic, and social basis for technological
change, but not the inventive act itself.

This work is, to an extent, an attempt to redress this neglect directly,
not only to understand better Edison and the nature of his contributions,
but also to suggest the extent to which invention—as an act and a process—may be scrutinized as a historical problem. It is fitting, therefore,
that the subject under study here is the very epitome of invention in the
cultural mythology of the twentieth century—the incandescent electric
light. The electric light is, of course, seen as the pinnacle of Edison's
inventive achievement. This is so in spite of the fact that other creations
may be said to have shown more originality (e.g., the phonograph), more
technical flair (e.g., the quadruplex telegraph), more persistence (e.g.,
the lead-acid storage battery), or a more ingenious combination of elements (e.g., the kinetoscope motion picture system). When it came
time, toward the end of Edison's life, for the world to pause and applaud
the Wizard's fantastic career, the occasion chosen was the "Golden Jubi-
lee of Light," October 21, 1929, the fiftieth anniversary of the date that,
Edison claimed, he invented the electric light. The primacy of the electric light in the constellation of Edison's inventions may be ascribed in
part to the size and influence of the industry that grew from the light and
its widespread use, in part to its omnipresence as a part of life and work

in the twentieth century, and in part to the almost spiritual significance that lamps and lighting have in human culture.

There are other reasons for making Edison's electric light the object of a study of invention. The general outlines of the invention of October 1879 are very familiar parts of each American's picture of his country's material progress in the nineteenth century, and yet the story that is such common knowledge is based much more on hearsay and foggy memories than on historical evidence. For the most part, it conforms with the traditional nineteenth-century image of invention, a story of persistence and sweat overcoming nature's intractability, of Edison and his faithful followers trying literally thousands of materials in the search for a suitable light bulb filament, guided by little more than the vision that nature had to provide some substance that would serve such a noble need. The electric light is, after all, the achievement most associated with Edison's famous aphorism defining invention as "99% perspiration." Little research is required to reveal how shallow and inaccurate an image this is of what was really going on at Menlo Park in 1878 and 1879.

There is, of course, a "revisionist" version of Edison's invention that tends, wrongly, to put the inventor into the mold of a twentieth-century manager of scientific and technical systems. This version, in its simplest form, would have us believe that, upon taking up the challenge of the electric light, Edison plotted out a research and development strategy encompassing available scientific knowledge about the subject as well as an understanding of the complex systems requirements of a complete electric light and power technology. This has served as a useful corrective to the naive popular view, but in fact reflects a naiveté all its own, casting Edison into a modern role he could never have assumed and making his achievement a far more straightforward and predictable act than it actually was. For Edison, the search for a practical incandescent light was a bold, even foolhardy, plunge into the unknown, guided at first more by overconfidence and a few half-baked ideas than by science or system. To suggest otherwise is to rob the inventive act of its human dimension, and thus to miss an understanding of the act itself.

Nor is it right to make Edison's invention simply one of many more or less equal steps in a long path leading from the first glimmerings of the theoretical possibility of electric lighting to the installation of the practical reality in homes, shops, and factories everywhere. The simple fact is that before Edison began his search in 1878, the world had nothing even resembling a practical electric lamp, and, when that search was largely over by the end of 1879 (and certainly by the time Edison's lamp was commercialized in 1882), the principles and form of the modern incandescent lighting system were established. It is not right either to make a great deal of the rivals Edison met in the field, whether in America or overseas, and to see in them equals in the enterprise. The evidence is simply not there to support the claim that any of these men possessed more than a portion of the whole that emerged from Menlo Park as the decade of the 1880s began. It will not even do to make much of the notion that one or two of these portions were, at critical junctures, pieced into the Edison system from the reports of his rivals—once again, the evidence is not there.

The invention of the electric light was a complex, human achievement, and we shall not understand it unless we fully appreciate that fact. This may seem to be an obvious truth, but little that has been written about the event has taken it into account. This invention, like most inventions, was the accomplishment of men guided largely by their common sense and their past experience, taking advantage of whatever knowledge and news should come their way, willing to try many things that didn't work, but knowing just how to learn from failures to build up gradually the base of facts, observations, and insights that allow the occasional lucky guess— some would call it inspiration—to effect success. There is clearly something to be said for trying to understand this process better, not just because it has been one of the most important agents for change in the last two centuries, but because it is a part of the human adventure.

This account of Edison's invention was shaped not only by the thematic goal discussed above, but also by a methodological goal of almost equal importance. The first goal is a truer and richer story based on a more faithful reading of the evidence, as opposed to the usual perpetuation or arbitrary inversion of myth. The second is an experiment in archival historiography. The experiment derives its rationale from the sponsor of this study, the U.S. National Park Service, which is the custodian of one of the richest and largest collections of historical technological documentation in the world, the archives at the Edison National Historic Site in West Orange, New Jersey. Adequately understanding and taking advantage of this unique historical resource has become a significant priority for the administrators and curators of the Site, as evidenced by its partial sponsorship of the Thomas A. Edison Papers project. While, as a comprehensive archival and publishing effort of more than twenty years planned duration, this project may be expected to provide the most thorough scrutiny of the archives' resources and potential, it is appropriate that other, more modest, attempts be made to explore their value. We seek, therefore, in this work to understand how to use a large and complete body of technical records to answer interesting historical questions.

As a glance at the accompanying references and bibliographical note will suggest, this study attempts to rely exclusively on the contemporary archival record of the activities surrounding the electric light's development from 1878 to 1882. There has been, over the years, a great deal written about Edison's premier invention and the circumstances surrounding it. Such writings began to appear only a very few years after the event and have continued up to this day, as exemplified by Robert Conot's well-received Edison biography of 1979, *A Streak of Luck,* and Thomas Hughes's study of electrification, *Networks of Power,* which appeared coincidentally with celebrations of a "centennial of light." The earliest works on the subject relied largely on the recollections of the still-living principals, for the archives were not available and Edison and most of his colleagues were usually quite ready to talk about their Menlo Park exploits. It took little time, however, for recollections to dim and for the complexity of events to be overshadowed in hindsight by the magnitude of the achievement. When, therefore, the stories and recollections of the pioneers, including Edison himself, are compared against the archival record, their completeness and accuracy are constantly found wanting. Later, more professional works, such as the biographies by Matthew Josephson (*Edison,* 1959) and Conot and more specialized studies by historians of science and technology, relied much more on the archival record and thus managed to avoid many of the more simplistic shortcomings of the earlier versions. Most of them, however, still relied in crucial places on recollections (most notoriously on Francis Jehl's *Menlo Park Reminiscences* of 1937–1941) or misinterpreted important technical elements of the record. In examining these accounts next to the Menlo Park notebooks, correspondence, and other documents, we have found none whose rendering of the events of 1878–1882 match our reading of the record. While, of course, some of these differences may be seen as simply matters of historical interpretation, we believe that many of them are due to differences in the degree to which the contemporary documentary record has been critically scrutinized.

This record has presented problems that have inhibited scholars from fully exploiting it. The first problem is the sheer size of the Edison archives. The document holdings at West Orange are said to contain more than three and a half million pages—as formidable a collection centered around the work of one man as exists anywhere. While there is no estimate of the size of the record that concerns the electric light alone, one can imagine that a four-year slice out of Edison's most productive years constitutes no small body of material. A better idea of just what this consists of may be found by a look at the Bibliographical Note following Chapter 8.

The second problem, related to the first, is organization. A fraction of the Edison archives is arranged by subject, but even that fractional arrangement was, at the time this study was undertaken, a somewhat unreliable and haphazard organization. Most of the relevant material on the electric light must be gleaned from amidst records dealing with other enterprises being carried on at Menlo Park. Some of the problems encountered here are also suggested in the bibliographical note. As wellmeaning and willing as the custodians of the Edison archives have been over the years, we can hardly wonder if they have been unable to fully allay the difficulties posed to scholars by such records.

Yet another problem is posed by the nature of the material itself. While there is nothing inherently incomprehensible about the notes, correspondence, and other papers generated in the Menlo Park laboratory, they are definitely not like the more literary records usually left by a political figure, writer, or businessman. They are the creations of men, immersed in the mechanical, electrical, and chemical knowledge of their day, attacking some very bedeviling technical problems. If the record they left behind them is generally without comprehensive explanations for their activities and ideas or interpretations of their abbreviated notes and scribblings, it should be no surprise. Indeed, those few documents that do seem to delineate more fully the ideas and purposes behind laboratory activity must be looked upon with suspicion, for they often turn out to be creations after the fact, put together for purposes of publicity or legal convenience. The papers actually produced in the course of laboratory activity are frequently but rough drawings of a new idea, quick calculations (with little or no labeling), lists of materials or devices, or descriptions of a laboratory procedure or observation without why or wherefore. All of this, it must also be remembered, is in the technical terminology of the nineteenth-century electrician or mechanic, an argot that can be as strange to the modern ear as the jargon of a computer programmer would be to one of the Menlo Park "gang." Therefore, most scholars and writers have understandably retreated to the much more straightforward accounts of the reminiscences for their image of the laboratory's workings and achievements.

We cannot claim that we have overcome these difficulties as completely as we would like, but we have made the effort to meet them head-on. The size of the record required the efforts of a full-time researcher for about eight months, spent searching out and noting down every relevant piece of data in the notebooks, correspondence files, scrapbooks, and other sources described in the bibliographical note. The organization of the electric light items necessitated a broad sweep through the documents of the 1878–1882 period. And the mass of technical arcana was dealt with forthrightly, with every effort made to comprehend and respect the technical milieu in which the men at Menlo Park worked. The extent to which we have in fact succeeded is, of course, a judgment we must leave to the reader.

Finally, a word should be said about illustrations. Because we are dealing here with mechanics, chemists, electricians, and other practical men, we must recognize that their most important form of communication was frequently not in words but in the quick sketch, the hasty set of figures, the finely detailed drawing, and the products of their workbench or laboratory table. Because history is largely a literary activity, often these records are translated here into descriptions, but such translation is never completely accurate and is frequently impossible. As a number of historians of technology have been at pains to point out, "nonverbal communication" is an essential part of the technical culture, and any student of that culture ignores this at great peril. We have attempted here to suggest the wealth of the nonverbal sources that are such an important part of the documentation of the electric light, but it must be remembered that our inclusions are but a fraction of the total nonverbal archival record.

CHAPTER 1

"A Big Bonanza"

In 1878, Thomas Edison was only thirty-one years old, but he had already produced enough significant inventions to credit a lifetime. The press recognized this achievement by calling him the "Wizard of Menlo Park." Beginning with his improved stock ticker of 1869, his contributions to telegraphy alone were enough to establish him as perhaps the premier electrical inventor of his day. His systems of automatic and multiplex telegraphy were not only technical marvels, but their possible economic significance made Edison's name as familiar to the financiers of Wall Street as it was to the followers of the technical and scientific press. Successful dealings with the telegraph empire builders of New York had given Edison the means to construct his unique laboratory in the New Jersey countryside. And there at Menlo Park he and a group of loyal co-workers operated a true "invention factory."

Soon after the lab was completed in the spring of 1876, Edison and his team moved beyond telegraphy. Their first important successes were in the field of telephony. Edison's carbon transmitter of 1877 was a crucial element in turning the experimental devices of Alexander Graham Bell and Elisha Gray into practical instruments for communication. The broad range of approaches Edison used in his inventive efforts sometimes yielded surprising results, as when in late 1877 experiments on repeating and recording devices for use with telegraphs or telephones resulted in the phonograph. The "talking machine" was surely Edison's most surprising invention. Despite the primitive quality of his tinfoil cylinder device, the public was agog at the machine. Most of the first months of 1878 were taken up by travel and demonstrations in response to the public clamor for showings of the phonograph. The "Wizard" became the object of enormous press attention, for hardly anything would seem to be beyond the capability of a man who could invent a machine that talked. Indeed, when New York's somewhat flamboyant *Daily Graphic* ran an April Fool's Day story headlined, "Edison Invents a Machine that Will Feed the Human Race," other newspapers repeated it as straight news.

Second Floor of the Menlo Park Laboratory, 1878.
This photograph shows some of the apparatus on the second floor of Edison's Menlo Park laboratory that made this the best equipped private laboratory in the United States.

For the first half of 1878 Edison basked in the spotlight. The surprise of the phonograph, along with the enthusiasm it generated from the public, turned his inventive energies away from their normally doggedly practical direction. He produced devices like the "aurophone" and the "telescopophone," both not very useful amplifying instruments. His observation of the changing resistivity of carbon under varying pressure led to the invention of the "tasimeter," intended as a supersensitive heat measuring device. All of these efforts were in part simply ways of showing off his inventive virtuosity, as well as a reaction to the lesson of the phonograph that even the most unlikely avenues of experimentation may yield wonderful discoveries.

The financial needs of the laboratory and its workers assured the continuance of more practical efforts. Much time and energy were devoted during these months to the further development of telephone components. The jumble of patents and conflicting business interests surrounding the technology of the telephone gave Edison and his backers the incentive to develop telephone devices that would complement the carbon transmitter and yet avoid the patents of Bell and others on receiving equipment. Later in 1878, the Menlo Park efforts would yield the chalk-

Edison and the Phonograph, 1878.
When Edison journeyed to Washington to exhibit his phonograph before the President, members of Congress, and the National Academy of Sciences, famed Civil War photographer Mathew Brady took this portrait of the inventor with his new invention.

drum telephone receiver, a clever device that was in many ways an improvement over other instruments but turned out to be impractical in broad application. Despite its obvious potential for lucrative profits and its technical similarity with telegraphy, the telephone already represented a crowded field, one that no longer held out the promise of quick breakthroughs. It should be no surprise, therefore, that in the middle of 1878 Edison was seeking fresher directions for his endeavors.

Edison's biographers describe his condition in the late spring of 1878 as "very tired and ill."[1] The never-ending round of public appearances to demonstrate the phonograph, claims and counterclaims surrounding his telephone inventions, and the constant grind of the lab had worn him down to the point where his need for a vacation was apparent to everyone. To the rescue came Professor George Barker of the University of Pennsylvania, who asked Edison to provide his tasimeter for use on the expedition to the Rockies being organized by Henry Draper for the purpose of observing a total eclipse of the sun due to occur on July 23. Barker accompanied his request with an invitation to Edison to join the Draper party, giving the inventor an opportunity not only to see the "wild West," but to spend several weeks in the close company of some of

Edison in the Spring of 1878.
This portrait shows how tired and ill the inventor appeared prior to his trip West.

America's most eminent scientists. When, on July 13, the scientists and their entourage departed New York for the long train ride west, Edison was with them.

The trip was clearly excellent tonic. While the tasimeter was of no value in its intended purpose (measuring the heat from the sun's corona), the escape from the East and the companionship of men like Barker and Draper restored Edison's energy and enthusiasm for new tasks. Edison's state of mind upon his return is reflected in some notes made by his chief assistant, Charles Batchelor, many years later:

> When he came back from this trip he told me of many projects to be worked up for future inventions, amongst them one for using the power of the falls for electricity & utilizing it in the mines for drills etc. He said he had talked a great deal with Prof. Barker who was his companion in a journey to the Pacific Coast after they had observed the eclipse in Rawlings. Prof. B had told him of some experiments he had seen at William Wallace's place at Ansonia, Ct. & wanted him to go up there & see them.[2]

The date of Edison's return was August 26, 1878. The researches that led to the invention of the incandescent lamp began the next day.[3]

It was almost two weeks later, however, before Edison threw himself

and his team wholeheartedly into electric lighting research. The push for this effort was provided by his visit to the factory of William Wallace. Professor Barker made all the arrangements, and the trip was made, in the company of Barker and Professor Charles Chandler of Columbia, on Sunday September 8. The firm of Wallace & Sons was the foremost brass and copper foundry in Connecticut and was known also for expert wire drawing. William Wallace himself had been experimenting with electricity for almost a decade and had built his first dynamo in 1874. He joined with the brilliant electrical inventor Moses Farmer and began manufacture of the Wallace-Farmer dynamo in 1875. Not many months before the visit from Edison, Wallace began development of an arc lighting system and the construction of a powerful electric motor-generator he called a "telemachon," (*tele*, from Greek, meaning distant) indicating that its primary purpose was the harnessing of electric power generated some distance away. A visit to Wallace's workshop in Ansonia was the best possible exposure to what America then had to offer in the infant field of electric light and power.

Such was Edison's notoriety that it was impossible for him to make such a trip without a newspaper reporter tagging along. The writer from the *New York Sun* was not disappointed, and he provided a lengthy description of Edison's reaction to what he saw:

Edison and the Scientists at Rawlins, Wyoming, 1878.
Edison accompanied a party of scientists to Rawlins to observe the July 29 solar eclipse. Among the party was George F. Barker, who encouraged Edison to work on electric lighting. Barker appears at the extreme left of the picture and Edison is second from the right.

Wallace Dynamo, 1878.
*Following his return from the West,
Edison visited William Wallace's shop
in Ansonia, Connecticut. A few
days later he ordered a small and a
large dynamo from the manufacturer
to use in experiments on electric
lighting. (Courtesy Smithsonian
Institution)*

Mr. Edison was enraptured. He fairly gloated over it. Then power
was applied to the telemachon, and eight electric lights were kept
ablaze at one time, each being equal to 4,000 candles, the subdivision of electric lights being a thing unknown to science. This
filled up Mr. Edison's cup of joy. He ran from the instruments to the
lights, and from the lights back to the instrument. He sprawled over
a table with the SIMPLICITY OF A CHILD, and made all kinds of
calculations. He calculated the power of the instrument and of the
lights, the probable loss of power in transmission, the amount of
coal the instrument would save in a day, a week, a month, a year,
and the result of such saving on manufacturing.[4]

The optimistic (or naive) *Sun* reporter then went on to describe the possibilities of harnessing Niagara Falls and distributing the resulting electric power throughout the United States. The final impetus for Edison's work on the electric light was provided not so much by the challenge of the light but by this vision of universal power through electricity.

While electric lighting was largely a new field for Edison it was by no means virgin territory. Ever since Humphry Davy in England had used his giant battery at the Royal Institution in 1808 to demonstrate how electricity could be made to produce light either by striking an arc between two conductors or by heating an infusible material to incandescence, the possibility of making a practical electric lamp had intrigued inventors and would-be inventors. The limited and expensive sources of current available before the 1860s, however, restricted serious efforts. Despite this, patents were taken out in several countries as early as the 1840s for both arc lights and incandescent devices.

The arc light was the subject of the most intensive work. Davy used two pieces of charcoal to show that a small gap in a circuit can be bridged by a strong current, producing a very bright, continuous arc. The technical problems presented by the arc light were straightforward: (1) producing electrodes for the points of the arc that would not burn up too rapidly from the arc's intense heat, and (2) finding a means of regulating the gap between the points so that the arc could be continuously sustained even while the electrodes were being shortened by the arc's destructive action. For more than forty years various devices were invented and developed to make the arc light practical. By 1878 the fundamental elements of arc light technology were well understood, and considerable progress had been made toward adapting the new light to appropriate uses. The light that William Wallace had on display at Ansonia was a typical example of what was then available—an electromagnetic regulator that held two carbon electrodes (plates in the case of the Wallace instrument, rods in most others) at the desired distance, producing a blindingly bright light upon the application of current. The arc light had found some use in lighting streets, public halls, and large stores, but was obviously not suitable for domestic lighting, where the desired light intensity was of the order of ten to twenty candlepower (the range of a gas light) rather than the thousands characteristic of the arc.

Whereas the arc light was beginning to find some applications in 1878, lighting by incandescence was far from being a practical technology. Davy had shown that an electric current could be used to heat a material to the point where it would glow. The basic problem was that almost all substances either oxidize or melt at temperatures sufficiently high to cause incandescence. One substance that would not melt at such high temperatures was carbon, but the ease with which carbon burns prevented experimenters from getting very far with it. The other popular substance for early efforts was platinum, whose resistance to oxidation was its primary attraction. Platinum's major drawback, besides its high cost, was the difficulty of raising its temperature to the point of incandescence without allowing it to heat up further, past the melting point (about 1770°C). All important efforts to make a workable incandescent lamp before 1878 used one or both of these materials.

As early as 1841, Frederick De Moleyns, an Englishman, received a British patent for an incandescent lamp using both carbon and platinum. In 1845 an American, J. W. Starr, not only patented two forms of incandescent lamp (one using platinum, the other carbon) but also traveled around England giving exhibitions and promoting his inventions. Starr's death at twenty-five the next year cut short his efforts which, while they impressed a number of well-informed British observers, were not in fact practical. For the next three decades a steady stream of devices flowed from the workshops of would-be inventors in Britain, America, and the Continent. Despite these efforts, the fundamental problem of the incandescent lamp in 1878—finding the means of heating an element to glowing without destroying it—was no closer to solution. In the words of the *Sun*'s reporter, "the sub-division of electric lights" was still "a thing unknown to science."

Upon his return from Ansonia Edison dived immediately into the task of producing a practical incandescent light. Notes from the Menlo Park laboratory made on September 9 and 10 referred to a number of platinum wire "burners" generally shaped into spirals. The notes in their rough way make it clear that Edison had been greatly excited by what he saw at Wallace's workshop. His excitement apparently stemmed not only from seeing what Wallace had accomplished but, more significantly, from perceiving how much had yet to be done. He described his feelings to the *Sun*'s reporter about a month later:

> [In Wallace's shop] I saw for the first time everything in practical operation. It was all before me. I saw the thing had not gone so far but that I had a chance. I saw that what had been done had never been made practically useful. The intense light had not been subdivided so that it could be brought into private houses.[5]

After only two or three days of experiments, Edison felt he was on to something—something big. On September 13 he wired Wallace urging him to send to Menlo Park one of his telemachons, "Hurry up the machine. I have struck a big bonanza."[6] Wallace wrote back the same day to assure Edison that a machine was on its way, adding "I truly hope you have struck a big bonanza." While the nature of this "bonanza" is not completely clear from the Menlo Park notes, there is enough evidence

Electric Light Aug 27. 1878

117

Copied from page 5 of Electric light records which 28 day of Sept. 1878
Wm Carman

Electric Light Experiment, August 27, 1878.
Edison's first electric light experiments following his return from Rawlins were similar to those he had performed the previous winter, when he brought materials such as boron and silicon to incandescence by placing them between the poles of a carbon arc.

for some surmises. On the same day he wired Wallace, Edison drafted his first caveat on electric lighting, "Caveat for Electric Light Spirals." Therein he wrote:

The object of this invention is to produce light for illuminating purposes by metals heated to incandescence by the passage of an

Telegraphy
and the Electric Light.

Edison began his electric light experiments confident that the solution to the problem lay in developing a regulator to prevent the filament from melting. His long experience with making and breaking electric circuits in telegraphy, particularly in multiple telegraphy, led him to assume that he could attack the light problem in a similar manner. The regulator used the expansion of metals or air heated by the electric current to divert current from the incandescing element in order to prevent it from destruction by overheating. A comparison of these two pages shows the similarities between Edison's electric light regulators and relay devices he used in multiple telegraphy.

Electric Light Draft Caveat #1, September 13, 1878.

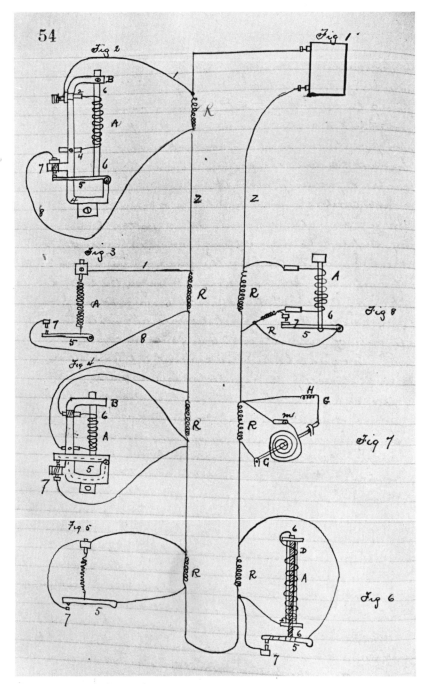

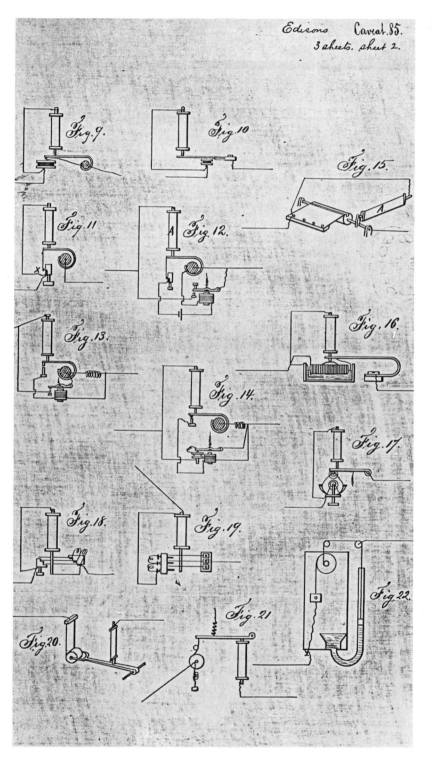

Electric Light, Edison's Caveat #85, October 25, 1878.

Multiplex Telegraph, Edison's Caveat #77, January 26, 1876.

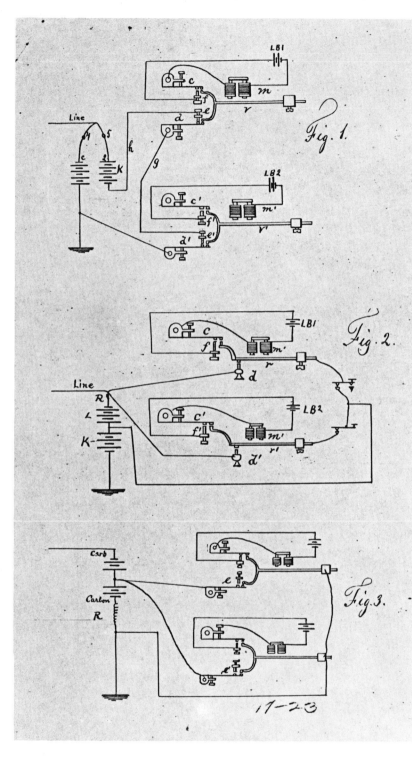

electrical current through them, a great number of pieces of such metals forming part of an electric circuit and distributed at various parts of the same. The invention consists in devices whereby the heat arising from the passage of such current is utilized to regulate the temperature of the incandescent metal which serves to give the light so that it is never allowed to reach its melting point, no matter how strong a current attempts to pass through.[7]

In the following pages of the caveat Edison described forty-four different regulator devices, all designed to "cause each spiral to automatically regulate its own temperature."[8] Most of these devices used the expansion of metal—either the incandescent spiral itself or another piece of metal nearby—to trigger an interruption or reduction of current when approaching the fusing point of the incandescent metal. The combinations of electromagnets, switches, resistance elements, and levers were clearly products of the telegraphic technology with which Edison was so familiar. Edison's confidence in his superior mastery of these mechanisms led him to believe that he could readily devise the stable and practical lamp that had eluded every previous inventor.

Notes from the Menlo Park laboratory show that the next few days were taken up with constructing some of the regulators specified in the caveat, and devising series circuits for them. Despite the fact that these notes do not reflect any signal success for these instruments, Edison's confidence grew into boastfulness. The *Sun* of September 16 carried a column headlined, "Edison's Newest Marvel. Sending Cheap Light, Heat, and Power by Electricity." "'I have it now!'" Edison was quoted as saying, "'and, singularly enough, I have obtained it through an entirely different process than that from which scientific men have ever sought to secure it.'"

They have all been working in the same groove, and when it is known how I have accomplished my object, everybody will wonder why they have never thought of it, it is so simple. When ten lights have been produced by a single electric machine, it has been thought to be a great triumph of scientific skill. With the process I have just discovered, I can produce a thousand—aye, ten thousand—from one machine. Indeed, the number may be said to be infinite. When the brilliancy and cheapness of the lights are made known to the public—which will be in a few weeks, or just as soon as I can thoroughly protect the process—illumination by carbureted hydrogen gas will be discarded.[9]

He then went on to describe how he would be able to light all of lower Manhattan with a 500-horse-power engine, using Wallace's dynamos, how underground wires would bring electricity into buildings, and how he intended to use existing gas burners and chandeliers as fixtures. The vision of a complete electric lighting system was clear in Edison's mind in the first days of working on the light.

The *Sun*'s story received considerable attention. Picked up by the *Philadelphia Bulletin,* it was seen by George Barker, who then wrote Edison regarding his "big strike in electric lighting." Barker remarked that he expected the news to have an effect on gas stocks and hoped Edison would be able to let him use some of his "new things" in lectures he was scheduled to give on the electric light that winter.[10] Other newspapers, including the *Chicago Tribune,* repeated the news. The most important impact, however, was in New York, where the story was read by some of the Wall Street moneymen who had already learned to be wary of Edison's technical genius. On September 17 Edison received a wire from his New York lawyer and friend, Grosvenor P. Lowrey, and some of Lowrey's associates requesting an urgent meeting. Shortly afterwards came a letter from Tracy R. Edson, an official of Western Union, who requested a meeting soon in his New York offices "in relation to your new discovery of which you spoke to me on Monday last."[11]

Thus began several weeks of negotiations between Edison, with Lowrey as his representative, and various financiers associated with the telegraph industry, gas companies, or both. Out of these talks was to come the Edison Electric Light Company, formed solely for the purpose of supporting Edison's experiments at Menlo Park and controlling the resulting patents.

The week (September 16–22) that began with the appearance of the *Sun*'s article was filled with the construction of experimental lamps. These were all based on the ideas put forth in the September 13 caveat—platinum wires or spirals in various forms of holders, with regulating switches triggered by the thermal expansion of the metal "burner" or an adjacent element. The belief that this was the right track was firmly held. Some of the sketches of instruments worked on that week show not only the regulating device, but also bases, stands, and connections.[12] On Sunday, the 22d, Edison wired his representative in Paris with the news that he would not be able to visit Europe soon: ". . . cannot come, have struck bonanza in Electric Light—indefinite subdivision of light. . . ."[13] He continued to receive inquiries sparked by the newspaper stories. George Bliss, who was in charge of promoting Edison's electric pen, wrote from Chicago to tell of the excitement the stories had stirred in that city and asking how much truth there was to all that he had heard. Edison's reply was a reflection of his confidence, "Say to Bliss [he instructed his secretary] Electric Light is OK. I have done it and it's only a question of economy."[14]

The "question of economy" was rapidly becoming a major concern, along with solving the continuing difficulties experienced with various versions of the regulator-burner. Notes made on September 20 show

Electric Light Subdivision Sept 17. 1878

131

Lamp and stand

also piece to go in front of this

copied from page 22 E. L. Record oct 1. 1878

Wm Carman

Lamp with Regulator and Base, September 17, 1878.
This carefully drawn lamp base indicates the extent to which Edison and his associates concerned themselves early in their research with the problem of attaching the lamp to fixtures in commercial applications.

calculations of the amounts of copper needed in various circumstances.[15] The quantity of copper needed to supply the huge currents that most assumed would be required by a large number of lights on a single circuit was one of the most glaring problems in any scheme for "subdividing the light." In the midst of all Edison's proposed regulators, there is little clue as to how he thought he would solve the distribution problem. Its significance was probably not apparent to him in those first few weeks of feverish excitement and boastful pronouncements. It was one thing to conceive of the electric light as part of an extensive and comprehensive light and power system, and another to perceive the technical requirements of such a system.

During the week following September 22, Edison drafted his second electric light caveat. The dozen or so devices described therein represented a hodgepodge of approaches. Self-regulating spirals of platinum were joined by arc light regulators, oxyhydrogen limelights fueled by electrolysis, sticks of carbon raised to incandescence in a vacuum, and devices that combined carbon, platinum, and other materials. Experimental notes from the last week in September show that efforts were still concentrated on improving the spiral-regulator lamp. A drawing of a lamp made on September 25 was accompanied by the comment, "We now have a perfectly regulating light spiral wound double to allow for expansion, . . . and when the spiral and Platina rod are the right size, this is a perfectly automatic cutoff. . . ."[16] The various experiments during this period, however, make it clear that the lamp was really far from "perfect." Various materials were tested in numerous ways; platinum continued to be favored, but iridium, platinum-iridium, ruthenium, and carbon also received some attention.[17] By the end of September, materials like chromium, aluminum, silicon, tungsten, molybdenum, palladium, and boron had been incorporated into parts of experimental lamps with few positive results. The obvious dissatisfaction with the behavior of platinum in his lamp was a clear sign that, for all the talk of a "perfectly regulating spiral," Edison was by no means certain that he was on the right track.

The pace of experimentation grew more intense as October rolled around. Some of the complexity of the task Edison had undertaken began to be reflected in the activity at Menlo Park. Each day his technicians, especially the talented mechanic John Kruesi, were instructed to make a variety of devices. The shape of the platinum element, the form of the regulator, the lamps' mechanical parts, and even the base and container of the devices were varied in countless ways. Experiments continued on alternative materials, although platinum remained the subject of most work, and titanium and manganese joined the already long list.

Edison's search for the ideal incandescent element was no secret. He wrote to Professor Barker in Philadelphia, for example, to get his opinion of the usefulness of titanium (to which Barker gave a somewhat negative reply).[18] A much more prescient communication, however, reached

Electric Light Draft Caveat #2, September 27, 1878.
These drawings illustrate the hodge-podge of filaments and regulator designs Edison considered.

Electric Light Division Sep 25, 1878

we now have a perfectly regulating light spiral wound double to allow for expansion, and the wire in middle same wire as the spiral only the spiral is flattened to make stronger, and allow more heating surface as the Platina rod expands or closes the cut off points at x and when the spiral and Platina rod are the right size. this is a perfectly automatic cut off; so that however much current you put on it will regulate before burning up although the contraction when cold has allowed the points to seperate 1/8 inch

Copied from page 36 E.L. Records Oct 1. 78
Wm Carman

A "perfectly regulating light spiral," September 25, 1878. *Although Charles Batchelor, who penned this notebook entry, expressed confidence in the new design, the lamp was far from self-regulating.*

Edison at about the same time (October 7) from Moses G. Farmer, who was himself experimenting on incandescent lamps. Farmer sent along a small bar of iridium which he thought superior to platinum as a light emitter. Farmer went on to say, however, that he thought none of these materials equaled carbon which "is the most promising—when sealed tightly from oxygen either in a vacuo or in nitrogen."[19] Farmer's iridium sample was probably much appreciated by Edison, who had already begun trying to use the material, but the hint on carbon was ignored, probably because his own brief experiments with the material showed that it was almost impossible to protect from combustion.

As important as the material problem was to Edison, after a month of intensive work it was still the mechanism of his light—the regulator—that commanded most of the effort of the lab. The third caveat for "electric light subdivision," prepared on October 3, and a fourth, written five days later, were both largely concerned with regulators. "Caveat No. 3," as it was headed, introduced the pneumatic regulator, which consisted of a platinum element enclosed in an airtight container equipped at one end with a diaphragm of unspecified material. As the element heated to incandescence, the air in the chamber expanded and pressed against the diaphragm, which was connected to a short-circuiting device adjusted to close when the air temperature reached a certain point, thus diverting current from the lamp element and preventing its destruction. This form of regulator was to receive more attention in the months ahead, eventually becoming the subject of patents. The fourth caveat largely covered a number of complex mechanical regulators, variations of earlier designs. The caveats also reveal a continuing concern as to the best shape for the platinum element, the October 3 document in particular depicting all kinds of spirals and bends. Various shapes were allowable because the pneumatic regulator did not depend upon the expansion of the incandescing element itself for regulation. The new type of regulator also directed attention to the behavior of the element in a closed container—a glass "bulb."

On October 10 notes were made for a series of experiments that differed significantly from most of those performed up to that time. Eight discrete experiments were described, mainly as part of a series to determine the relationship between radiating surface areas, temperatures, and light emission of incandescing elements. The tests marked the first notable departure from the somewhat haphazard construction of various devices toward a more systematic experimental approach. These experiments were not particularly well designed. Even the best two or three involved only a couple of trials; for example, one compared the light output of two platinum strips of the same resistance and length but different widths, one-eighth and one-quarter inch. Edison's explanation for this effort was, "The idea is to ascertain if we do not gain increased light by increased surface without alteration of resistance and also to ascertain if the radiation nullifies the effect of increased surface."[20] By mid-October the Menlo Park experimenters were beginning to sense how much they had yet to learn before the electric light could be a practical reality.

It was against the background of this frenetic experimenting that the complex business arrangements were completed for supporting the increasingly expensive researches and eventually for exploiting the resulting patents. In late September some members of the New York legal and financial world had attempted to get commitments from Edison regarding the financing of the electric light research and the disposition of rights to his invention. Three individuals carried on the most active correspondence: Grosvenor P. Lowrey, who had acted as Edison's attorney since 1877; Western Union's Tracy R. Edson; and Hamilton McK. Twombly, another Western Union official and the son-in-law of William

Electric Light Draft Caveat #3, Pneumatic Regulator, October 1, 1878.

Edison introduced the pneumatic regulator in early October. It used heated air to short circuit the lamp and divert current from the incandescing element to prevent its destruction. This form of regulator received much attention in the months ahead, eventually becoming the subject of patents.

T. A. EDISON.

Caveat No 3

Menlo Park, N. J., Oct 1   187

Regulating by air expansion

Oct 1, 1878.

ok

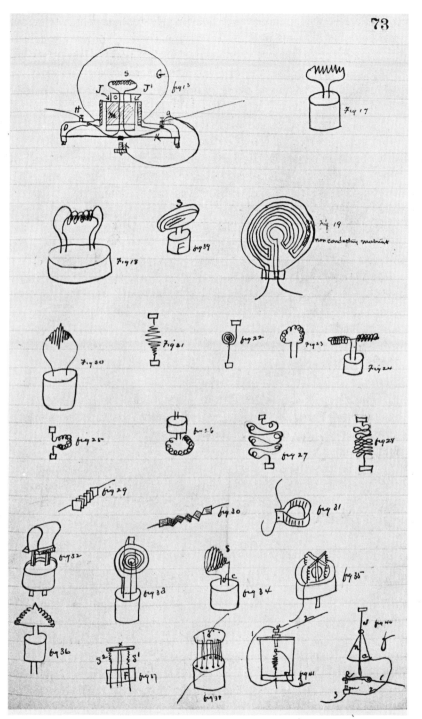

Electric Light Draft Caveat #3, Forms of Burners, October 3, 1878.

As important as the regulator was in preventing the destruction of the incandescing element, Edison also explored many forms for the burner to make it inherently more durable.

H. Vanderbilt, whose money was heavily invested not only in telegraph companies but also gas utilities. From an early date, Lowrey sought to represent Edison's financial interests in electrical lighting. During the last week of September, the Western Union financiers attempted to get Edison's ear but he was able to avoid them, protesting that no time could be spared from the Menlo Park lab. By early October, Lowrey had the negotiations under control. He worked out an arrangement with the Vanderbilt interests to establish a stock company with a capital of $300,000, half to be represented by Edison's interest in his invention and half by money supplied to Edison by Vanderbilt, Twombly, Edson, and other shareholders. At this point, Edison was quite content to let Lowrey make whatever arrangements he thought best, writing on October 3, "Friend Lowrey: Go ahead. I shall agree to nothing, promise nothing and say nothing to any person, leaving the whole matter to you. All I want at present is to be provided with funds to push the light rapidly."[21]

By this time it had become clear to Edison that he would require resources beyond what he already had at Menlo Park. He was eager for the financial support that Lowrey's arrangements would give him and apparently had no misgivings about sharing control of his invention. On October 5 he wrote enthusiastically to one of his European representatives, Theodore Puskas:

> The electric light is going to be a great success. I have something *entirely new*. Wm. H. Vanderbilt and friends have taken it in this country and on Monday next advance $50,000 to conduct experiments.
>
> I retain ½ of the capital stock of the Co. they are to form and also receive a royalty of $30,000 yearly if it proves more economical than gas, which *I am certain it will do*. Vanderbilt is the largest gas stock owner in America.[22]

However, to another European representative, George Gouraud, Edison advised a bit of caution. "Say nothing publicly about light," he wired on October 8, "I have only correct principle. Requires six months to work up details. Gas men here hedged by going in with me."[23] It soon became clear to Edison that, while his financing would give him the resources his experiments needed, it also increased the pressure on him to produce viable results quickly. For the next year the efforts at Menlo Park had to be directed not only to the creation of the practical light Edison wanted, but also to producing evidence that could convince investors of progress toward a profitable invention. The two goals turned out to be frequently incompatible.

The need for additional money in early October was the result of not only the seemingly endless experiments with the light and its various forms of regulators, but also a growing awareness of the need to develop other components of an electric lighting system. The most crucial of these was the generator. It is important to remember that the experiments begun in the fall of 1878 marked Edison's first real encounter with the production of electric current by mechanical means. All his previous electrical work—with telegraphs, telephones, and a host of minor devices—was satisfactorily served by batteries. There was never any doubt that a practical electric light could be economical only if powered by an electromagnetic generator. The beginning of serious work on electric lighting at Menlo Park coincided with the acquisition of the telemachon that had so impressed Edison on his visit to William Wallace's Ansonia workshop. For the first few weeks Wallace's machine appeared sufficient, but this judgment did not last long. On October 4 an exchange of telegrams between Edison and Wallace concerned the cost to Edison of a second Wallace machine, perhaps a more powerful one. Edison ordered one immediately at Wallace's confidential discount price of $750.[24] At about the same time, Edison began shopping around for alternative machines, sending an inquiry to the firm of Arnoux and Hochhausen and placing an order on October 10 with the Newark supply house of Condit, Hanson and Van Winkle for a "dynamo electric machine" made by Edward Weston.[25] The next day Edison received, apparently in response to an earlier inquiry, information from the Telegraph Supply Company of Cleveland, Ohio on the generator and arc light of Charles Brush.

The ultimate sign of Edison's dissatisfaction with what he had available was the beginning of efforts to design his own generator. The first fruit of this was the construction of a "tuning fork magneto," a device of little practical use which Edison nevertheless saw fit to patent (U.S. Patent 218,166; granted Aug. 5, 1879). After the middle of October 1878 the construction of a suitable generator was to be one of the major tasks at Menlo Park, sometimes overshadowing work on the light itself. From this point on, there could be little question that Edison perceived his task in terms that went beyond a workable light and included the system needed to make it practical.

While the work at Menlo Park gradually revealed the complexities and difficulties that lay in the path toward the electric light, public excitement over what was promised from Edison's laboratory was unabated. Indeed, in early October the news from America caused a panic in gas shares in London, where Edison's reputation for wizardry was unequaled. Throughout the month, Edison received sometimes frantic letters and cables from George Gouraud in London, who claimed that he was besieged by capitalists. "If I had had my wits about me," he wrote, "when your telegram came announcing your discovery in this connection I might have made you a clean million as it played the very devil with stocks all over the country."[26] The next week Gouraud cabled to say that the "tremendous excitement" continued and that "there never was a time so favourable to the launching of a large company than the present for the Edison

Telegraphy and the Generator.
The design of Edison's first patented generator (U.S. Patent 218,166) evolved from work with tuning forks in his acoustic telegraph experiments in 1876. These experiments led him to devise various electric motors incorporating a tuning fork design, such as U.S. Patent 200,032 for synchronous movements of telegraph apparatus. The similarity of motor and dynamo design made the tuning fork generator an obvious first step for him.

Tuning Fork Motor, 1878.

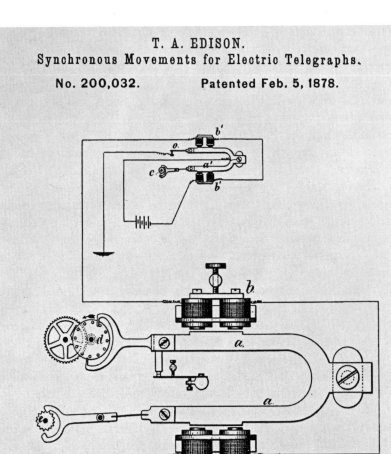

Tuning Fork Generator, 1879.

T. A. EDISON.
Magneto Electric Machine.

No. 218,166.   Patented Aug. 5, 1879.

Electric Light Company."[27] Edison resisted Gouraud's entreaties, referring all business matters to Lowrey, who in turn preferred to move conservatively and only with the approval of the New York financiers he already had behind him.

The newspaper stories in both America and Europe continued to be enthusiastic. The lack of public demonstrations after a month of promises was said to be due to the wait for patent protection. Reporters described their visits to Menlo Park where Edison, busy as he was, seemed always ready to greet them. The visit to the laboratory was usually highlighted by a demonstration of the light. Rhapsodized a reporter from the *New York Sun*:

> There was the light, clear, cold, and beautiful. The intense brightness was gone. There was nothing irritating to the eye. The mechanism was so simple and perfect that it explained itself. The strip of platinum that acted as burner did not burn. It was incandescent. It threw off a light pure and white. It was set in a gallows-like frame, but it glowed with the phosphorescent effulgence of the star Altaire. . . . It seemed perfect.[28]

The intensive publicity continued until almost the end of the month. By that time there was little to tell the public but that they would have to wait—a few technical details, completion of the patent procedures, some financial arrangements, and all would be ready.

At the same time, however, Edison's financial backers in New York were seeking somewhat more concrete assurances. Lowrey wired to Menlo Park to suggest that it was important that Edison be prepared to demonstrate his invention soon.[29] The reply said it was only a matter of days before some kind of showing would be ready. Nonetheless, the delays were making some backers nervous, and approaches were made to the rival inventor, William Sawyer, who, with the backing of Albon Man, claimed to have devised a workable carbon incandescent lamp. When Edison received word of the discussions with Sawyer, he was furious, revealing perhaps some of his own anxiety. His secretary, Stockton Griffin, wrote to Grosvenor Lowrey, describing the scene:

> Upon arriving at the Park yesterday, I spoke to Mr. Edison regarding our conversation about the Sawyer-Mann [Man] electric light, being careful not to say anything beyond what you told me. I was astonished at the manner in which Mr. Edison received the information. He was visibly agitated and said it was the old story, i.e. lack of confidence. The same experiences which he had had with the telephone, and in fact with all of his successful inventions, was being re-enacted. He also referred to the telephone being loaded with useless encumbrances and remarked that if he had a voice in the matter the electric light should not be so treated. No combina-

tions, no consolidations for him. . . . He said that it was to be expected that everyone who had been working in this direction, or had any knowledge of the subject would immediately set up their claims upon ascertaining that his system was likely to be perfect. All this he anticipated, but he had no fears of the result knowing that the line he was developing was entirely original and out of the rut.[30]

Lowrey hastened to reassure Edison:

Do not give yourself the slightest uneasiness from anything you hear from me or anybody else, about other people's efforts or inventions concerning electric light. My confidence in you as an infallible certain man of science is absolutely complete. . . .[31]

Despite Edison's attitude, the backers of the Edison Electric Light Company insisted on some insurance for their investment. At the same time that approaches were being made to Sawyer, Lowrey wrote to Edison that the Company's Board of Directors wanted to hire someone "to examine the state of the art" in electric lighting, and went on to ask if Moses Farmer would not be a good expert for the job.[32] Edison himself asked his patent attorney for a list of all U.S. patents on the electric light.[33] He also suggested to the Light Company directors that Howard R. Butler of the Gold & Stock Telegraph Company be hired to make a thorough search of the patent and technical literature to compile a survey of the work that had been done on electric lighting. The directors agreed a few days later to pay for Butler's work, and the survey promptly began.[34] At about the same time Butler was asked to take time from his job at Gold & Stock, he agreed to take into his office a young student of physics, Francis R. Upton, freshly returned from study in Germany. Upton wrote to his mother on November 7 that Butler did not yet have anything definite for him to do.[35] This changed quickly when the request from Edison arrived, and Upton was assigned the search Edison's backers wanted. Thus was recruited the most important single collaborator Edison was to have in the coming year's work.

As November wore on, it became increasingly evident to Edison and to at least some of his associates that the invention of the electric light was far from the simple matter many had supposed it to be. Edison began to show an active interest in the new effort to investigate all that had been written on electric lighting. On November 12 he wired Butler in New York that he wanted to meet with Upton, presumably to get a progress report on the literature search.[36] He also began work on the construction of a brick machine shop—a much more substantial building than previous Menlo Park facilities. The new building would require a good portion of

Francis Upton, c. 1882.
(Courtesy Smithsonian Institution)

the first $50,000 he had gotten from the Light Company, but he felt he needed a place with "all the means to set up and test more deliberately every point of the electric light, so as to be able to meet and answer or obviate every objection before showing the light to the public or offering it for sale either in this country or in Europe."[37] In this same letter to his agent in Paris, Edison characterized his efforts in revealing terms:

> Before I have done with it I mean to succeed. I have the right principle and am on the right track, but time, hard work and some good luck are necessary, too. It has been just so in all my inventions. The first step is an intuition, and comes with a burst, then difficulties arise—this thing gives out and then that—"Bugs"—as such little faults and difficulties are called—show themselves and

months of intense watching, study and labor are requisite before commercial success—or failure—is certainly reached.[38]

A keen awareness was emerging at Menlo Park of the magnitude of the "bugs" that had to be worked out to make a practical electric light, but there was no question that Edison "meant to succeed."

After mid-November, however, there was a clear change in the path being charted to success. Edison's interest in what Upton was finding in his literature search was one clue to this change. On November 22 Upton wrote a ten-page summary of his findings to date, referring to a number of patents that he thought Edison should know about. He indicated his desire to learn more from Edison and added some assurances, "I want you to explain me your lamp and tell me exactly what you claim. . . . I feel sure that the total you have is new, no matter if the parts have been used before."[39]

In late November, Edison put in an order to the New York firm of Willmer & Rogers for subscriptions to a long list of technical journals, including the *Gas Light Journal, Metallurgical Review, English Coal Gas Journal,* and others that would allow him to keep abreast of developments in lighting, metallurgical chemistry, and other fields of interest.[40] Finally, in late November, Edison instituted a new procedure for recording experimental researches in the lab, which began the extraordinary series of laboratory notebooks that document the work from that time on. The long, hard months of "watching, study and labor"—the months of inventive perspiration—were also beginning.

When Edison launched himself and his laboratory on the already popular search for a practical, useful means to "subdivide the electric light," the chief characteristic of his search was confidence. In those first weeks, at the end of the summer of 1878, he saw the problem of the electric light as electromechanical, and no one in the world had more self-assurance that he could solve problems of this kind than Edison. And, indeed, the world at large shared this confidence. The problem of subdividing the light was, after all, a well-known, oft-attacked challenge, one that had already stumped some very good electrical inventors. But once the Wizard of Menlo Park had announced that the essential solution was in fact simple—one that readily fit into that technical domain of which he had proven himself to be master—then at least the lay press was happy to share that perception. For all the bravado that rings through Edison's pronouncements of those first weeks, the fact is he genuinely believed that he had indeed solved the problem of the electric light.

The central technical feature of his solution was the principle of the self-regulating element. The key to a successful light, Edison believed, was the devising of a reliable mechanism to maintain a balance between the power needed to make the lamp element (entirely or largely of plati-

Precision Machine Shop, Menlo Park Laboratory, 1878.

This is the original precision machine shop located in the rear of the Menlo Park laboratory. It was very similar to those Edison was familiar with as a telegraph manufacturer. While the shop was ideal for making small precision parts for telegraphs or even electric lamps, it was not suitable for building large-scale machines such as dynamos.

Large Machine Shop, Menlo Park Laboratory, 1879.

With funds from the newly organized Edison Electric Light Company, Edison was able to build this large machine shop in the winter of 1878–1879. It allowed him to build large devices, such as the dynamo depicted in this Scientific American *engraving, and to move quickly ahead of his rivals, who had fewer resources, in developing a successful electric light system.*

num) incandescent and the power that would cause the element to melt or otherwise destroy itself. Edison rapidly came to the conclusion, which he held for the better part of the next year, that the incandescing element could be harnessed to a negative-feedback regulator that would help to maintain this balance. In those first weeks, the problem of the electric light was perceived almost solely in terms of making the regulator work. Other aspects were simply not seen as important.

Concentration on the regulator did not mean that Edison was unaware of the ultimate need for a well-developed system to make his light work. Indeed, everyone who thought for even a moment about the problem of electric lighting recognized the necessity of linking several discrete technical elements into a coordinated system. Edison's perception of the systemic nature of the lighting problem was not especially sophisticated. Models of technical systems surrounded every late nineteenth century inventor—the telegraph, gaslight, and arc light systems were only the most obvious examples. That an electric light would have to have practical power generation and supply networks behind it was not a novel concept. The more elaborate appreciation of the technical requirements and possibilities of an electric light (and power) system that was to

The Menlo Park Laboratory Complex, 1881.
This painting, by R. F. Outcault, in the winter of 1880–1881, shows the laboratory complex as it appeared during the height of work on electric lighting. The long wooden building in the center is the main laboratory, the small building in the right foreground is the brick library, and the large building in the rear is the brick machine shop.

emerge from the Menlo Park labors of the next year was hardly present at all in the beginning. Edison did make the point that his electric light would operate in just the fashion of the gaslight, but this was more boast than model. And the image that was sometimes drawn of Niagara supplying all of America with light was clearly more dream than plan.

As the fall of 1878 wore on, the spirit behind the work at Menlo Park changed. The serious difficulties encountered in the design and perfection of the self-regulating lamp showed that even the purely electromechanical challenges were beyond those Edison had so handily overcome before. Furthermore, the larger complexities of the electric light began slowly to unfold. The complexity of the system itself—the need for a generator precisely designed for its task, for understanding and creating circuits soundly based on electrical science as well as suited to the kind of light people wanted, and for ultimately creating a myriad of components to make the system reliable and efficient—only gradually dawned upon the workers at Menlo Park. The seeking-out of financial backing and the construction of new laboratory facilities were clues to this growing awareness. Recognition of the need to absorb the technical literature and even to draw upon scientific expertise—a new experience for an inventor who was not unhappy in his characterization as "wizard"—was another indication of how the perception of the task had changed. The invention of a practical electric lighting system would prove to require months of difficult, frustrating, and consuming labor— and a bit of luck.

CHAPTER 2

"The Throes of Invention"

Up to the end of 1878, Edison's attack on the problem of "subdividing the light" was really little different from that of would-be inventors who had preceded him or of rivals who were then in the midst of their own efforts. What was to distinguish Edison's work in the coming months (and years) was the wealth of men, equipment, and facilities that he could mobilize for the campaign. No other inventor in the nineteenth century had at his disposal what Edison had—a team of skilled and intelligent co-workers armed with every instrument, tool, or material they required and dedicated to the accomplishment of whatever end Edison set out for them. As the search for a practical light moved into 1879, the scope of effort that the Menlo Park team and laboratory made possible began to have an impact.

Between 1876, when Edison moved to Menlo Park, and the years 1881–82, when he began phasing out the laboratory there, the number of men working in the little group of buildings hard by the Pennsylvania Railroad tracks varied considerably. There were, however, never more than a half-dozen central figures in the lab's work. These individuals differed enormously in background, training, and skills but possessed in common, at least while they were at Menlo Park, an extraordinary loyalty to Edison and faith in what they could accomplish under his guidance. Their loyalty and confidence was perhaps the most important factor in making the Menlo Park laboratory an effective and productive cooperative enterprise. Even if it had been conceivable, the corporate structure of the twentieth-century industrial research laboratory was not necessary in a setting built around the inspiration and leadership of one man. Nonetheless, Menlo Park was also a place that brought out the most important talents of the individuals there—talents that were themselves critical to the successful pursuit of invention.

Unquestionably, the chief among Edison's co-workers was the English-

Menlo Park Laboratory Staff, 1879.

Edison, in the center without a vest, is surrounded by his staff on the steps of the laboratory. Included are Charles Batchelor, the bearded figure over Edison's left shoulder, Francis Jehl, immediately to the left of Batchelor, and Francis Upton, to Jehl's left.

born mechanic Charles Batchelor. Raised in Manchester and receiving most of his training in textile mills, Batchelor came to America at age twenty-two to help a Newark factory with its installation of machinery. When he shortly thereafter joined Edison at his Newark shop, he quickly became an indispensable part of the operation. Batchelor was particularly valued for the fineness of his handiwork and the painstaking care and patience he put into all he did. It would perhaps be going too far to call him a foil to Edison, but much of his value clearly lay in the extent to which his methodical manner balanced Edison's more rough-and-ready tendencies. More than anyone else, it was Batchelor who was to be found by Edison's side at the laboratory workbench.

John Kruesi was another valuable Menlo Park hand whose association with Edison dated from Newark days. Kruesi was a master machinist whose mechanical skills reflected his Swiss background. It was his Menlo Park machine shop that was responsible for turning the roughly sketched ideas of Edison and others into real constructions of wood, metal, and wire. If the devices that emerged didn't work, it was because they were

Charles Batchelor, c. 1882.
(Courtesy Smithsonian Institution)

bad ideas, not because they were badly made. And when the ideas were good, as in the case of the phonograph, the product of Kruesi's shop would prove it. The fact that the Menlo Park laboratory possessed the mechanical capability of a first-rate machine shop was no small element in its success. Not only did the quality of the shop's output give good designs their best chance of working, but it also allowed the rapid testing and elimination of poor concepts. It was advantages like this that set Menlo Park apart from the environment of any other inventor in the world.

Edison also owed much to the other members of his team, even if their contributions were not as singular as those of the lab's most skilled or knowledgeable individuals. John and Fred Ott were mechanics who worked for Edison from Newark days until well after Menlo Park. Samuel D. Mott was a draughtsman with an artistic flair who was responsible for the attractively detailed drawings of laboratory devices scattered throughout the Menlo Park notebooks. Martin Force was another whose name recurs throughout the laboratory records. He came to Menlo Park with little special training but, under Edison, became a valued laboratory

John Kruesi, c. 1882.

assistant. The crew of Menlo Park also included chemists, metalworkers, bookkeepers, secretaries, and general laboratory helpers—all dedicated to carrying out Edison's designs.[1]

For years Edison had surrounded himself with skilled craftsmen whose abilities with machines and materials made up for his own limitations. In the search for the electric light, however, a new capability was called for, hitherto absent from the Menlo Park lab, and it was embodied in the person of Francis R. Upton. A graduate of Bowdoin College, Maine and a recipient of postgraduate training in physics at Princeton and under Hermann von Helmholtz at Berlin, Upton brought with him a sophistication in physical theory and scientific practice that had been lacking. That lack had hardly been noticed before, for scientific training—apart from a familiarity with the rudiments of electricity and chemistry—had never seemed relevant to Menlo Park's mission. Indeed, when Edison invited Upton to come to Menlo Park upon completion of the literature and patent survey, he was probably more attracted by the young scholar's diligence and eagerness to please than by his academic credentials. Edison had required no persuading to go along with the literature search on which his backers had insisted in November, and the (apparently) impressive manner in which Upton carried out the search, along with his obvious intelligence, was reason enough to suppose he would be useful in the laboratory.

The manuscript records from Menlo Park rarely speak directly of day-to-day activities in the laboratory, although much can be inferred from notebook entries. For a more vivid picture of laboratory life and routine, the best source is the popular press. Much has been written about Edison's relations with journalists, and there is little question that a key element in the forming of Edison's reputation and popular following was the almost instinctive way in which he cultivated reporters—tolerating their intrusions when he would stand no others—and the grateful and eager way in which the newsmen reciprocated the favor by chronicling the miracles of Menlo Park with uncritical and bright-eyed wonder.[2] In light of this relationship, and the mixture of ego and reportorial license that taints its product, care must be taken in using the newspapers' descriptions and pronouncements as testimony for what actually went on at Menlo Park. Nonetheless, nothing gives the flavor of life and routine there better than the journalists' eyewitness accounts, as in this example from the *New York Herald* of January 17, 1879:

The ordinary rules of industry seem to be reversed at Menlo Park. Edison and his numerous assistants turn night into day and day into night. At six o'clock in the evening the machinists and electricians assemble in the laboratory. Edison is already present, attired in a suit of blue flannel, with hair uncombed and straggling over his eyes, a silk handkerchief around his neck, his hands and face somewhat begrimed and his whole air that of a man with a purpose and indifferent to everything save that purpose. By a quarter past six the quiet laboratory has become transformed into a hive of industry. The hum of machinery drowns all other sounds and each man is at his particular post. Some are drawing out curiously shaped wire so delicate that it would seem an unwary touch would demolish them. Others are vigorously filing on queer looking pieces of brass; others are adjusting little globular shaped contrivances before them. Every man seems to be engaged at something different from that occupying the attention of his fellow workman. Edison himself flits about, first to one bench, then to another, examining here, instructing there; at one place drawing out new fancied designs, at another earnestly watching the progress of some experiment. Sometimes he hastily leaves the busy throng of workmen and for an hour or more is seen by no one. Where he is the general body of assistants do not know or ask, but his few principal men are aware that in a quiet corner upstairs in the old workshop, with a single light to dispel the darkness around, sits the inventor, with pencil and paper, drawing, figuring, pondering. In these moments he is rarely disturbed. If any important question of construction arises on which his advice is necessary the workmen wait. Sometimes they wait for hours in idleness, but at the laboratory such idleness is considered far more profitable than any interference with the inventor while he is in the throes of invention. . . ."

As the research at Menlo Park moved on, a constant consideration was financing and the confidence of the Electric Light Company backers. It continued to be Grosvenor Lowrey's task to keep relations between Edison and the financiers on an even keel. Lowrey had what must have often seemed a thankless job, calming the nerves of the Wall Street men whose consternation at the difference between the promises of imminent breakthroughs and the obvious struggling of the laboratory efforts was quite real (and justified). At the same time, the faithful lawyer had to soothe the easily ruffled ego of the inventor, who bristled whenever doubts of his eventual success were voiced. While Lowrey's diplomacy was important throughout the period of the electric light's invention and development, it was never tested more than in the late fall and early winter of 1878–79. Men who had been lured into backing Edison by the assurances of quick success were understandably upset over the lack of visible progress. The financiers had to learn much the same lesson as Edison—that the incandescent light was a complex invention, not to be achieved quickly or simply.

Lowrey held Edison's financing together through an adroit combination of candor and bluff. Toward the end of November he suggested to Edison that a visit to the laboratory by the Wall Street people would be a good idea, despite the lack of a successful light to show them: "It is all the better that they should see the rubbish and rejected devices of one sort and another. Their appreciation thereby becomes more intelligent."[3] Perhaps because he was sensitive about the problems he was running into, Edison did not always welcome these visits, so Lowrey found himself delicately balancing the concern of the New Yorkers against the qualms in the laboratory. Early in December, he expressed his difficulties to Stockton Griffin, Edison's secretary:

I have put them [a group of investors] off once or twice, telling them that we thus take up time which is of great value, but I do not like to repeat this too often. Edison must therefore allow this to be added to the interruptions, and after next Monday's visit I believe everybody will have seen what is to be seen, and he will then be left free to pursue his studies without further interruption, and strongly supported by the sympathy and confidence of his friends and associates.[4]

Following the visit that resulted from Lowrey's arrangements, he wrote Edison:

> The visit yesterday was productive, I think, of solid good results. Our friends had their imaginations somewhat tempered, but their judgements are instructed, and we now have to deal with an intelligent comprehension of things as they are, which makes both your part and mine much easier. They realize now that you are doing a man's work upon a great problem and they think you have got the jug by the handle with a reasonable probability of carrying it safely to the well and bringing it back full.[5]

Lowrey, in fact, continued to have to deal with grumbling investors, but he was always careful in his communications with Edison. After a couple of the New Yorkers returned from a disappointing trip to Menlo Park in late December, Lowrey described their impressions to Edison: "In addition to not finding you, they say that the general dilapidation, ruin and havoc of moving caused the electric light to look very small, and that it looked rather as if you were getting ready to have an auction."[6] He then suggested that Edison should allow the Electric Light Company men to keep a closer watch on expenses, though he hastened to assure him that whatever was absolutely necessary would be paid for. Although he certainly did not say so in his letters to Menlo Park, Lowrey was probably getting nervous about the appetite for funds being shown by the effort. He hence sought to bring in major additional sources of money, the most important of which was Drexel, Morgan & Company. By negotiating on the basis of the foreign rights to Edison's light patents, Lowrey managed to get the crucial extra backing Edison needed as the new year began.

Edison's efforts were indeed beginning to be expensive. He wrote in early January of 1879 to his Paris agent: "The fund I have here is very rapidly exhausted as it is very expensive experimenting. I bought last week $3000 worth of copper rods alone, and it will require $18,000 worth of copper to light the whole of Menlo Park ½ mile radius."[7] Expenditures like these, coming amidst reports from the laboratory that the platinum lamp was unworkable, were enough to make even the sturdiest investor hesitate. It is even more remarkable, therefore, that J. Pierpont Morgan and his associates should choose then to give Edison further backing. In a letter to Edison dated January 25, Lowrey described the remarkable scene in which Morgan affirmed his confidence in Edison's efforts despite setbacks:

I saw . . . that these gentlemen were likely, in a stress, to turn out—as I always supposed they would—not to be very easily frightened away from a thing they once made up their mind to. All they, or I, shall ask from you is to give confidence for confidence. Express yourself, especially when you come to a difficulty, freely. You naturally, having an experience of difficulties and of the overcoming of them, in your line (which none of the rest of us can have), may feel that it would be prejudicial, sometimes, to let us see how great your difficulties are, lest we, being without your experience in succeeding, might lose courage at the wrong time. This will be true sometimes of all people, but every active mind greatly interested in a particular subject works in its own way when a difficulty is presented in finding out the causes, and reasoning against the probability of their being insuperable; and with our friends I think that would be the result in almost every instance where you yourself should show that you still believe in a possible success.[8]

In accounting for Edison's accomplishments, not least of the advantages he enjoyed over rival inventors was the extraordinary trust placed in him by some of the greatest financial figures of his time. To be sure, this confidence was earned only after years of astonishing achievements, from which the financiers learned never to take lightly Edison's works or claims. The backing given him in 1878 and 1879, however, marked a new stage in the relationship between money and invention in America—a glimpse of the era in which giant corporations would routinely expect technical expertise, backed by science and laboratory resources, to turn out newer and better products for profit.[9]

As 1878 drew to a close, the work in the laboratory gradually became more systematic. The nature of the important technical problems was becoming better understood, and the laboratory's efforts reflected this by becoming more focused on crucial details. Equally important, the design of laboratory experiments was changing. No longer were most activities narrowly directed toward making better devices, in hopes that a breakthrough would result, but whole series of experiments were carried out to increase the understanding of the materials being worked on and the forces being manipulated. Edison and his co-workers had absorbed the lesson that the route to success lay through a return to basic principles. Too many new phenomena and poorly understood effects were being uncovered to expect their goal to be achieved by simply redesigning devices until they worked. Backed by his assistants and by a well-equipped and well-financed laboratory, Edison could afford, for perhaps the first time in his life, to seek a deeper understanding of the scientific and technical foundations upon which he must build.

Machine for Insulating Filaments, December 4, 1878.
This drawing by Batchelor shows one of many devices designed to assist in the production of experimental lamps. Insulators allowed spirals to be wound tightly without short-circuiting.

In late November and through December little progress was made in the area that had been the central concern, the lamp itself. Batchelor continued efforts to make platinum spirals that would hold up under the conditions required for incandescence. A complex device was designed for packing spirals with chalk—one approach to making the spirals as tight as possible while still insulating the strands.[10] A few experiments were run to get a better sense of how the platinum wire expanded and contracted during a heating and cooling cycle,[11] but the laboratory notebooks for the most part show that Edison's attention in this period had been diverted elsewhere, primarily to the problem of the generator.

Edison had devoted some effort to the mechanical production of electricity as early as mid-October, when he came up with his first designs for the tuning-fork generator. The tuning-fork design continued to appeal to Edison on into December, when he filed for his patent on the idea. For a few weeks Batchelor continued to attempt improvements in the tuning-fork machine,[12] but by mid-December Edison was ready to move beyond it to work that continued steadily for months until he finally had what he needed. This work started with further testing of the most popular generators then available, leading gradually to a rejection of them all. Edison explained his experiments to a *New York Sun* reporter:

> I am all right on my lamp. I don't care anything more about it. Every bit of heat is utilized to producing light as far as art will allow. The theoretical and practical results are perfectly satisfactory. My point now is the generator. The Wallace machine gives me three lights, each equal to a gas light, to a one-horse power. I feel sure that I can get six with an improved machine. Probably I can get more. Now, to make my grand practical experiment here in lighting Menlo Park, I should have to use twenty or thirty Wallace machines. They would cost me from $30,000 to $40,000. They would be useless afterward, for I know that I can make a generator of double their power. So I shall postpone the experiment until I find the machine that will give the greatest amount of electricity per horse power.[13]

When William Wallace read this, he was understandably upset and protested to Edison that it was unfair to judge his arc light machine by its performance with incandescent lamps. If Edison would only tell him the sort of device he wanted, Wallace was sure he could supply it.[14] But Edison could not yet tell anyone what kind of generator he wanted—he simply didn't know.

Intensive efforts were made to acquire other machines. A Siemens dynamo was ordered through Wallace, and when Edison had trouble getting a Gramme machine, he wired to Prof. Henry Morton at Stevens Institute in Hoboken and to George Barker, asking to borrow one.[15] When Barker suggested that he get one at Princeton instead, Edison was a step ahead of him, Batchelor having already started experiments with the fairly small Princeton device. Efforts to obtain one of Charles Brush's generators were not so successful[16] but Edison was apparently not concerned, having come rapidly to the conclusion that the Gramme machine was the most advanced one available and hence worthy of the most intensive experimentation. Several laboratory notebooks are full of the results of tests on the Gramme machine and others, as well as sketches of numerous possible armature windings, magnet orientations, and commutator designs. Edison, Batchelor, and Upton spent most of the latter part of December 1878 engaged in this work at the expense of all other activity, including work on the lamp. It is a significant aspect of Edison's method that, even in this period of learning the fundamentals of generators, possible new designs are interspersed through all the notes with little or no theoretical justification.

The range of tests being made on the dynamos reflected Edison's perception that he needed to acquire a basic understanding of how the machines worked and what principles might guide their modification. While some observations dealt with mundane considerations of friction and vibration, others were clearly attempts to deduce broadly useful guidelines. For example, Upton noted at one point Edison's conclusion that "When the internal resistance of the magnet equals external the best effort is obtained. Ohm's law applies to magnetism."[17] The work, always combining tests of the machines with modifications and new designs, was often intense. Batchelor noted several elaborate experiments on Christmas Day, and his drawings of new armature configurations and other modifications show that experiments continued without break through the last week of 1878 and into the new year. On January 2, 1879, Batchelor wrote in his own notebook, "Edison's Magneto Electric Mach. Have begun to make a practical working machine after a few weeks hard study on magneto electric principles."[18] Other laboratory notebooks, however, indicate no breakthrough in early January but simply a shift in emphasis from tests of existing devices to constructing new ones. It would be several months yet before a truly satisfactory generator emerged from the Menlo Park lab.

Gramme Generator from *Revue Industrielle,* May 2, 1877.
The Gramme generator was the first commercially successful dynamo and, by 1878, was being used to power arc light systems. Edison considered it the most advanced machine available and conducted many experiments to both understand how it worked and learn the general principles of dynamo design. (Courtesy Smithsonian Institution)

During late December 1878 all work on the lamp had ceased, making way for the intensive generator experiments. Despite the fact that the generator effort was still inconclusive, lamp-related work resumed shortly after the beginning of the new year. The new experiments reflected the changed spirit now permeating the work at Menlo Park. Not only the experimental style but the kinds of questions being asked were different. The lack of fundamental knowledge about what was going on in the lamps, about the root cause of their failure, was finally seen as a major obstacle. Instead of being characterized by the construction of numerous lamp prototypes, each intended as potentially patentable models, experiments were now carefully conceived tests and investigations of lamp materials, resulting in some remarkable observations on the behavior of materials under the conditions required for incandescence. On January 3 Upton began almost two weeks of tests on platinum-iridium and iron wires, recording their deformation and changes in resistance when heated to incandescence.[19] Edison himself began similar observa-

Experiments with Gramme Generator.

These are two of many notebook entries describing experiments with the Gramme generator. In the first, Charles Batchelor illustrates the direction of current in the Gramme armature. In analyzing his experiment, Batchelor considered the current in the generator's armature analogous to that produced by batteries. A similar analogy was used by Francis Upton in drawings that liken the poles of a Gramme armature to a series of the copper and zinc electrodes commonly used in batteries.

Gramme Generator Experiments, December 9, 1878.

6

Gramme Machine   Dec. 9 1878

Imagine a N pole passing into a solenoid and no S to follow the current would be all in one direction.

As you pass the coil along the bars the current will be all same direction from X to X. It would be the same thing as putting one bar in coil half way and then turning it round & putting it out of the other end

In Gramme armature the top half has a current running in one direction and bottom half a current in the other direction these meet at o o and the product is same as two batteries for quantity

tions of the kinds of changes platinum and platinum-iridium wires underwent when heated.[20] Batchelor soon joined in the careful testing, measuring, weighing, and inspecting that marked Edison's intensive drive to learn more about what was actually happening when he put his chosen "burner" candidates under the extreme conditions of prolonged and intense heat associated with incandescence.[21]

The most remarkable example of the new experimental approach was a series of observations that Edison recorded between January 19 and 29 under the heading, "Experiments with Platina & Platinum-Iridium alloys 20 per cent Ir—at the incandescent point with galvanic battery to determine any changes that may take place."[22] Here Edison systematically wrote down every pertinent detail about the behavior of the wires he was studying, adding additional materials to his original list as he went on. The resulting notes are distinguished by the obvious care with which

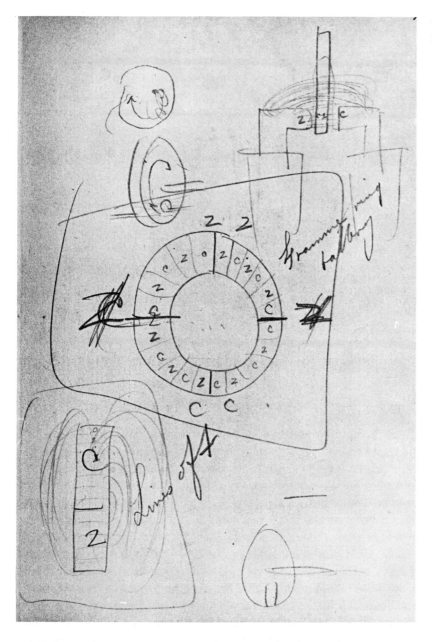

Gramme Generator Experiments, c. December 1879–January 1880.

Edison composed them and by the meticulous drawings he interspersed throughout, aided by an instrument heretofore little used at Menlo Park, a microscope. Thus, he noted not only the gross deformations of platinum wire as it was heated, but also the resulting changes in the structure of the material, the appearance of cracks and globules, and subtle differences in how the wires broke after they had been heated. Wires of platinum and different iridium alloys of platinum were compared not only on the basis of their different compositions, but also as they differed in

**Magnet Experiments,
December 25, 1878.**
Many magnet experiments were conducted to learn the principles behind electromagnetic generation of current. In this notebook entry Charles Batchelor describes an experiment to determine the best relationship between the armature and a field magnet.

gauge (thickness) and from supplier to supplier. Considerable effort went into determining not only physical but also chemical changes in the wires, careful weighing and flame tests being part of the procedure.

After a few days of working on platinum alloys, Edison shifted his attention to other metals, sometimes using foil strips instead of wires. Palladium behaved much like platinum, and microscope observations revealed not only the cracks found earlier but also what appeared to be a bubble. Gold proved impossible to bring to incandescence. Ruthenium, iridium, rhodium, and iridosmine (a naturally occurring iridium-osmium alloy) were tested, with results that were more colorful than encouraging. Tests of platinum-iridium were varied by using a Gramme machine in place of the customary battery, and by heating with an oxyhydrogen flame rather than an electric current. Small losses in weight were taken as evidence of some volatilization of the platinum, but the effect was not consistent. A wide variety of metals was tested for fusibility in an oxyhydrogen flame, and when Edison observed that nickel appeared to be as infusible as platinum, he determined to pursue further tests of the more common and cheaper metal. On January 23 Edison wrote:

We take a piece of nickel and roll it out, cut a narrow strip and pass a current through it and very strange to say it becomes brilliantly

25   Jan 2nd 1879

*Edisons
Magneto Electric Mach.   ChatBatchelor*

Have begun to make a practical working machine after a few weeks hard Study on magneto electric principles.

*Feb 15 1879
we have not been able to take off these currents as yet as the pull on the shell has ruptured the wire and we were obliged to cut it up*

Edison Magneto-Electric Generator, January 2, 1879.
After several weeks of intense testing of machines with various modifications and new designs, work began on constructing the first generator designed by Edison.

incandescent *without fusing.* I think it nearly if not equal to platinum. It slowly oxidizes but we shall prevent this by sealing the burner. When it does fuse it acts like the Pt-Ir 20 pc alloy—it remains hard when incandescent. It is very probable that absolute chemically pure nickel will have a very much higher fusing point than the sample we have which is probably only commercial. This is a great discovery for the electric light—in the way of economy. [23]

Obviously, if nickel were equal to or better than platinum in the regulator-type lamp, it would greatly improve the economics of light.

So excited was Edison about his nickel experiments that word spread rapidly. A January 25 letter to Edison from Grosvenor Lowrey in New York quoted recent visitors to Menlo Park as reporting that Edison had already abandoned platinum for nickel. Rather than receiving this as good news, however, Lowrey admonished Edison to "be sure you are right about nickel and everything else before having anybody know about it," pointing out that the report of such a major change in the light's composition undermined investors' confidence in what had been achieved to date. [24] Even before he had received Lowrey's warning, however, Edison's enthusiasm for the cheaper metal had dissipated. Further tests of nickel on January 24 had revealed considerable oxidation even after a short pe-

Platinum Experiments, January 1879.

By the beginning of 1879, difficulties with incandescent elements were apparent, and a more systematic approach was adopted. The most re-markable example of this new ap-proach was a series of experiments on the properties of various metal burn-ers recorded by Edison between Janu-ary 19 and 29.

riod (ten minutes) of heating to a yellow heat. While investigations of the behavior of nickel under incandescence continued, tests of other materi-als resumed, and nickel was soon relegated to the long list of momen-tarily promising substances that did not withstand further testing. The wide-ranging experiments continued for a few more days. For example, Edison rigged up an electric arc powered by a Gramme generator to heat his test substances, only to have to cease after a day of looking at

It *has* not anything like the strength of platinum when at the same temperature.

When the Pt-Ir wire is allowed to remain incandescent for 30 minutes a remarkable change is effected It stretches, grows smaller in diameter (ie) from 004 To 003. & breaks upon examination under the microscope it has this appearance

Platina wire thus

If the platina wire is only brought to incandescen for say 2 minutes it shews these cracks

**Platinum Experiments,
January 1879.**
*Edison's experiments with platinum
wire were aided by the use of a heretofore little-used instrument, the
microscope. Edison was then able to
see not only gross deformations of the
platinum wire as it was heated but
also changes in structure, cracks and
globules, and subtle differences in
how the heated wire broke.*

the glaringly bright arc caused his eyes to "suffer the pains of hell." By the end of January, however, Edison seemed to have satisfied himself that no metal other than platinum would do for a lamp element, and thus further work would have to concentrate on improving the capability of platinum or a platinum alloy to withstand the extreme conditions imposed by the lamp.

The January experiments did not reveal any obvious way to protect

**Platinum Experiments,
January 1879.**

*The platinum experiments suggested
to Edison that the burner problems
lay less in the composition of the
metal than in the environment in
which it was heated. Absorption of
gases was seen as particularly important and led Edison to focus his attention on minimizing its effects by
placing the burner in a vacuum.*

I notice what appears to be an air

bubble   thus.

I noticed on taking away from the plane
that under the microscope a large piece
of some salt, or organic matter was
attached to it — This might have caused
the fusion by reducing the temperature
at that point.

I again put the paladium to incandescu
for 15 minutes, under the microscope the
whole surface is crystaline, shewing the
crystals perfect. these are true crystals
& look like Antimony. they do not

the platinum from cracking and eventual disintegration. Observations did
seem to suggest, however, if only vaguely, that the problem lay less in
the composition of the metal than in the environment in which it was
heated. The sources of this inference are not clear from the laboratory
notes, but there is no question that it was made. The most obvious
means for altering that environment and reducing its deleterious influences on incandescing platinum was to minimize the environment it-

self—in other words, to produce a vacuum. Hence, in late January, Edison and his colleagues turned their attention for the first time toward creating a lamp in an evacuated glass envelope—a "light bulb." There was nothing new about this concept, for the very earliest attempts to produce a practical incandescent light had involved preventing oxidation of the glowing element by enclosing it in a vacuum. At the outset of his work, however, Edison had rejected this approach, relying instead on a nonoxidizing element (platinum) which needed protection from melting rather than oxidation. The initial confidence at Menlo Park had been based on the belief that protection against melting by an electromechanical feedback device would be adequate, and that the insuperable hurdle faced by previous inventors—providing a permanent high vacuum for each light—could be entirely avoided. Now it seemed that it could not.

Edison's acceptance of the need to join the search for a vacuum lamp was, in part, a recognition of how deeply committed he had become to producing a successful lamp. Too much of his reputation and too much money was now at stake to allow him to pull back, even at the point where it looked as if he might have to tread the same fruitless path broken by others before him. If Edison was daunted by this forced shift in direction, the notes from the laboratory do not show it. Some of Edison's confidence in entering this hitherto avoided territory may have been due to the knowledge that improved vacuum pumps were available to make his work easier than that of his predecessors. Most notable was the mercury pump first described by Hermann Sprengel in 1865. Sprengel's pump, a modification of one devised by Heinrich Geissler, allowed the repeated evacuation of a space with such efficiency that previously unheard-of vacuums had been obtained in laboratory applications. On January 22 Edison cabled both Professor Barker in Philadelphia and Professor Morton in Hoboken asking to borrow a Sprengel pump "for a few days."[25] Neither man could provide one, and Edison had to do without (sketchy notes made on January 23 suggest that an attempt was made to design a mercury pump for Kruesi to construct, but give no result).

Lack of a Sprengel pump did not stop the pursuit of vacuum experiments, and on February 3 the first vacuum lamp was sketched by Batchelor and Kruesi.[26] Soon thereafter began a series of tests on the behavior of platinum in a vacuum using a mechanical pump. These tests were apparently guided by a hypothesis derived from earlier experiments and carefully recorded by Upton in a notebook on February 4: "An explanation of the changes wh[ich] occur in Pt. may be the following. The Pt. absorbs an enormous amount of H gas which is given off at high temperatures."[27] Edison confirmed this to his satisfaction in the tests that followed, writing "I think from our experiments that the melting point is determined greatly by the amount of gas within the pores of the metal which by expansion disrupts the metal and makes it fuse easier."[28] Be-

Geissler Pump.

This is one of many clippings in the Menlo Park scrapbooks that indicate the interest of the staff in the technical literature on vacuum pumps. The best then available was the Sprengel pump, a modification of one developed by Heinrich Geissler.

represents this machine as constructed by Alvergniat. It consists of a vertical tube, which serves as a barometric tube, and communicates at the bottom, by means of a caoutchouc tube, with a globe which serves as the cistern.

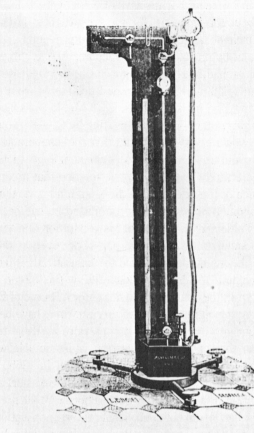

FIG. 146. Geissler's Machine.

" At the top of the tube is a three-way stop-cock, by which communication can be established either with the receiver to the left, or with a funnel to the right, which latter has an ordinary stop-cock at the bottom. By means of another stop-cock on the left, communication with the receiver can be opened or closed. These stop-cocks are made entirely of glass. The machine works in the following manner :—Communication being established with the funnel, the globe which serves as cistern is raised, and placed, as shown in the figure, at a higher level than the stop-cock of the funnel. By the law of equilibrium in communicating vessels, the mercury fills the barometric tube, the neck of the funnel, and part of the funnel itself. If the communication between the funnel and tube be now stopped, and the globe lowered, a Torricellian vacuum is produced in the upper part of the vertical tube.

" Communication is now opened with the receiver ; the air rushes into the vacuum, and the column of mercury falls a little. Communication is now stopped between the tube and receiver, and opened between the tube and the funnel, the simple stop-cock of the funnel being, however, left shut. If at this moment the globe is replaced in the position shown in the figure, the air endeavours to escape by the funnel, and it is easy to allow it to do so. Thus, a part of the air of the receiver has been removed, and the apparatus is in the same position as at the beginning. The operation described is equivalent to a stroke of the piston in the ordinary machine, and this process must be repeated till the receiver be exhausted.

" As the only mechanical parts of this machine are glass stop-cocks, which are now executed with great perfection, it is capable of giving good results. With dry mercury a vacuum of $\frac{1}{250}$th of an inch may very easily be obtained. The working of the machine, however, is inconvenient, and becomes exceedingly laborious when the receiver is large. It is therefore employed directly only for producing a vacuum in very small vessels ; when the spaces to be exhausted of air are at all large, the operation is begun with the ordinary machine, and the mercurial air-pump is only employed to render the vacuum thus obtained more perfect."

yond this, Edison believed he had found a solution to the problem of absorbed gases: "By gradually increasing the heat the gas gradually comes out of the metal without disrupting or cracking it. Roughly speaking, I think that if the melting point of platina in the air by suddenly bringing to incandescence is 2000°C then its melting point is raised to at least 5000°C by subjecting it to the process of occluding the gas by heat in a vacuum."[29] So taken was he by this effect that for the next few days he proceeded to investigate the effect of a vacuum on the melting of other metals under incandescence. He tested steel, iron, and magnesium wires, yielding interesting results but little useful information. Upton continued the experiments, trying out not only different metals but also a variety of coatings on wire spirals.[30] Several lamps were designed at this time with spirals of wire wrapped around bobbins of compressed lime or similar minerals, the bobbins serving to support the tightly wound spirals on an insulated base.[31] During the month of February the laboratory notebooks were filled with numerous designs for vacuum lamps featuring spirals of wire with and without bobbins. Confidence was growing that the adoption of the vacuum represented a major breakthrough.

As February drew to a close, it did indeed seem as though an important new stage had been reached. On March 1 Edison prepared applications for an American patent that would protect not only the new vacuum techniques but also an approach he had only hinted at earlier—lamps of high resistance.[32] Later that spring he consolidated specifications into a wide-ranging application for a British patent that incorporated a number of other advances. So clearly does the draft of the British specifications (recorded in a laboratory notebook[33]) spell out what Edison believed he had accomplished that it deserves an extended look. (Page numbers given for the following quotations refer to the notebook.) He began by describing his discoveries of what happened to platinum when it was heated for a period of time and then allowed to cool: "The metal is found to be ruptured and under the microscope there is revealed myriads of cracks in every direction, many of which are seen to reach nearly to the centre of the wire" [pp. 31–33]. He then stated that he had also found that platinum lost weight when heated and that the combination of the cracking and weight loss made ordinary platinum unsuitable for use in an incandescent lamp. These patterns were caused, he explained, by "the gases contained both in the physical pores and also in the mass [of the platinum]" [p. 41]. Driving out the gases by heating the wire spirals in a vacuum was then described as the means for remedying these defects. He specified that a Sprengel pump should be used, and "if the mercury pump be worked continuously and the temperature of the spiral raised at intervals of 10 or 15 minutes until it attains to vivid incandescence and the bulb be then sealed, we then have the metallic wire in a state heretofore unknown, for it may have its temperature raised to the most dazzling incandescence, emitting a light of 25 standard candles" [p. 51]. Fi-

Vacuum Lamp Experiments, February 1879.

Lack of a Sprengel mercury pump did not stop vacuum experiments. During the first week of February experiments like this one were performed with lamps evacuated by a mechanical air pump.

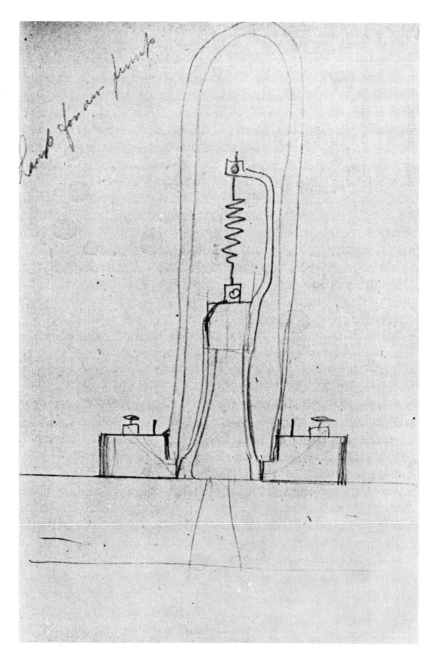

nally, he claimed, if the wire is coated with magnesium oxide, an even brighter and more durable light is obtained.

Having described fully the significantly improved behavior of platinum (or platinum-iridium alloy) in a vacuum, the specifications proceeded to give details of the lamp itself and the system in which it was to be placed. Here, for the first time, Edison clearly spelled out his other great discovery—the importance of high resistance:

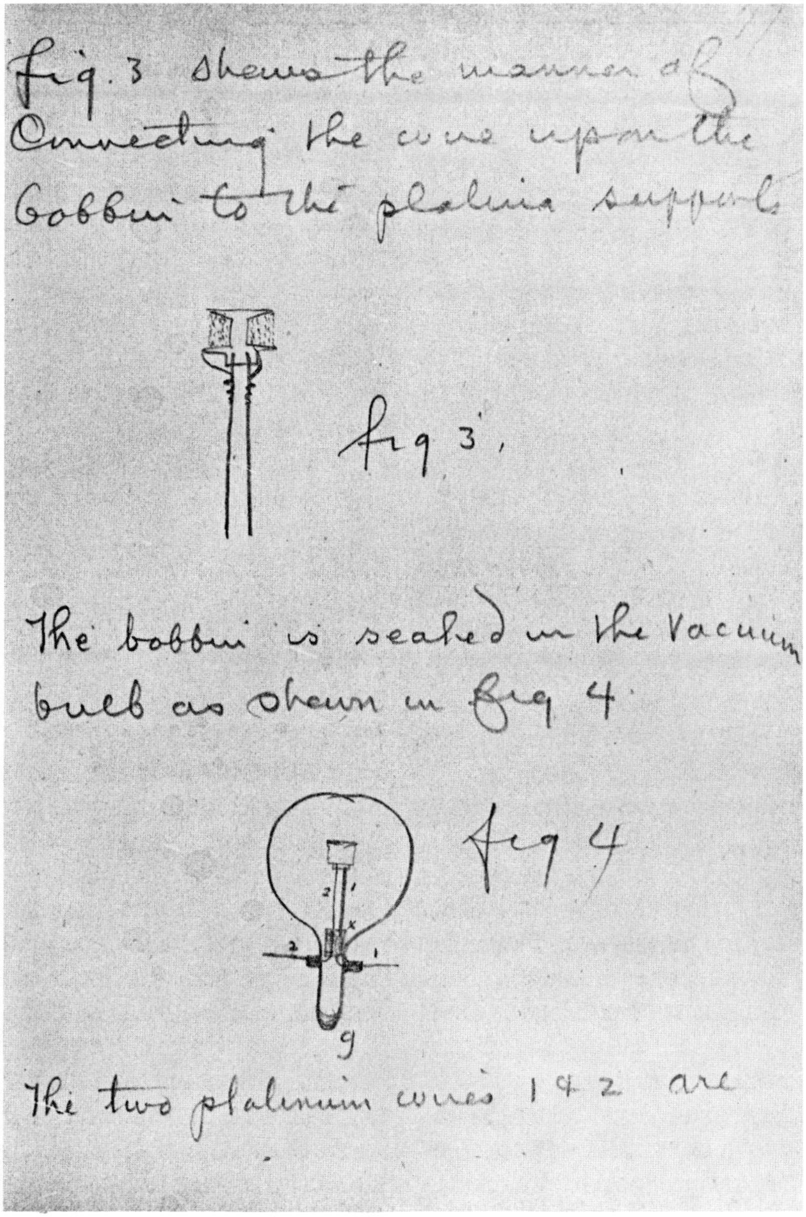

fig. 3 shews the manner of
Connecting the ccre upon the
Bobbin to the platina support

fig 3.

The bobbin is seated in the vacuum
bulb as shewn in fig 4.

fig 4

The two platinum ccres 1 & 2 are

**Bobbin Insulators,
Spring 1879.**
In drafting his provisional British patent specification, Edison described burners made of platinum wire rolled on bobbins made of infusible oxides. Edison felt that coating the wire with magnesium oxide made it harder and more satisfactory.

To obtain practically several hundred electric lights each equal to the ordinary gas jet upon one circuit it is essential for many reason(s) both on the score of economy, facility, and reliability to place them all in multiple arc and to prevent the combined resistances of several hundred lamps from falling to such a low point as to require a main conductor of immense dimensions with low resistance and generating machines of corresponding character. It is essential to

reverse the present universal practice of having lamps which have
but one or two ohms resistance and construct lamps which shall
have when giving the proper light [resistance] of several hundred
ohms [pp. 63–67]

He then went on to explain how the energy consumed by each lamp (for
a given quantity of light) was proportional to the lamp's radiating surface,
not its resistance; thus, high resistance lamps would require no more energy to operate than low resistance ones, but would allow the use of a
"conductor of very moderate dimension. . . . In practice, a resistance of
200 to 300 ohms in the burner will be sufficient" [p. 69].

There were hints in some of the tests carried out in the months previous to the British application that the men at Menlo Park were dimly
aware of the significance of lamp resistance for the practicality of an electric lighting system. For example, in late November 1878 Batchelor remarked in notes on the system of nitrogen-filled carbon lamps recently
introduced by William Sawyer and Albon Man, "If worked for quantity it
would want enormous large conductors owing to the small resistance in
each carbon."[34] (The Sawyer-Man carbon rods were reported to have a
resistance of one ohm.) In an evaluation of the Werdermann system,
compiled at this time, Batchelor again noted the difficulty posed by low
resistance lamps.[35] In mid-December Upton noted the fact that the energy consumed by high resistance lamps in a multiple (parallel) circuit
was the same as that of low resistance lamps in series.[36] Measurements
of resistance—of lamps, circuits, and generators—became common features of tests during this period, but it was not until late February 1879
that the critical importance of high resistance lamps was clearly stated,
as in the specifications for the British patent application. The recognition
of this importance was a major breakthrough for Edison. Understanding
that a viable "multiple-arc" system of lighting could not be built around
the kind of low-resistance devices other inventors had turned out was a
crucial step forward. It was one of the things that began to distinguish
the work at Menlo Park from what had gone before. In the future, when
Edison had to defend the patents protecting his invention, his most reliable claim to novelty was, in fact, the central place of high-resistance
lamps in his system.

Although this grasp of the high resistance principle was of crucial importance, it alone could not yield a workable light. In Edison's patent
drafts during this period, clear explanations of the uses of a vacuum and
high resistance elements are followed by descriptions of regulators that
were nothing less than nightmarish in their complexity. The proposed
British specifications described in detail an electromechanical regulator
for each vacuum bulb that was a maze of wires, springs, magnets, and
shafts. Edison explained that "the main object is to produce even illumination . . . not particularly to prevent the fusion of the incandescent con-

Resistance and the Sawyer-Man System, November 28, 1878. *Recognition of the importance of high lamp resistance was a major breakthrough for Edison. Charles Batchelor's discussion of the Sawyer-Man light in this notebook entry suggests that the Menlo Park laboratory staff understood this very early.*

ductor."[37] His insistence that the vacuum treatment of the platinum made a protective regulator unnecessary is hard to reconcile with the obvious complication that his regulator introduced into an otherwise elegantly simple lamp or with the fact that specifications for a March 1, 1879, high-resistance American patent application made no such disclaimer.[38] The American application, however, described a pneumatic regulator that involved enclosing the vacuum bulb in a second glass container equipped with a diaphragm that would, as the temperature of the

Resistance and the Werder-mann System, November 30, 1878.

Entries in Charles Batchelor's notebook on parallel connection of Werdermann arc lamps are further evidence of the thought given by the Menlo Park staff to the importance of resistance for subdividing the lamp.

lamp climbed beyond desired limits, "disconnect the lamp from the circuit, where it remains until the temperature is reduced to the normal condition." These regulating devices represented no real progress over those made in the fall of 1878, and the continued need for them was perhaps the primary reason that Francis Upton wrote home to his father, "The light does not yet shine as bright as I wish it might, but I am not despairing [sic] at all but that success will come sometime in the future."[39]

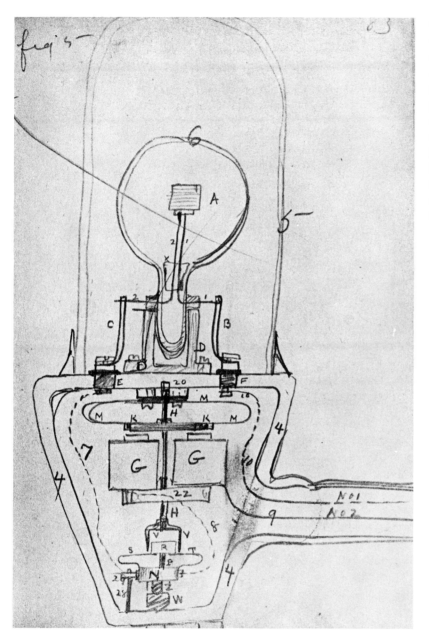

Complicated Lamp Regulator, Spring 1879.

Although Edison felt that vacuum treatment of the platinum wire would make a regulator unnecessary, his patent applications continued to describe regulators that were nothing less than nightmarish in their complexity. This drawing from his proposed British patent shows a vacuum bulb with an electromechanical regulator that was a maze of wires, springs, magnets, and shafts.

Pneumatic Lamp Regulator, March 1, 1879.

This pneumatic regulator from Edison's American patent specification was less complicated than its British cousin but represented no real improvement over regulators proposed in the fall of 1878.

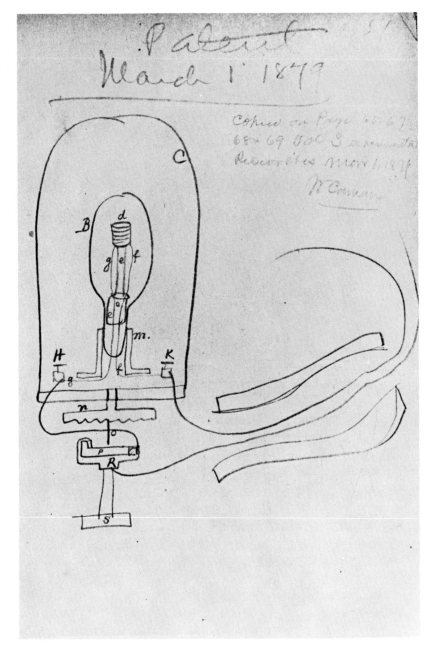

The Search for a Vacuum

On January 19, 1879, Edison became aware that gases were entrapped in platinum and suspected they might play a major role in its mechanical and electrical properties. His January experiments with platinum and other metals indicated that, at high temperatures, these gases caused bubbles and cracks.[1] The experiments also suggested that the problem lay less in the metal's composition than in the environment. The most obvious means for minimizing the problem was to create a vacuum. In late January the first concerted effort began toward producing an evacuated lamp.

There was nothing new about this concept, for the very earliest attempts to produce a practical incandescent light had involved enclosing the glowing element in a vacuum to prevent oxidation. Edison's own first experiments in January 1877 used a manually actuated piston pump of a type he had acquired as early as 1874.[2] With it he could achieve a vacuum on the order of 2.5 millimeters mercury, or about 0.003 atmosphere.[3] (An atmosphere is a pressure of approximately 15 pounds per square inch, the usual air pressure at sea level, which will support a column of

mercury approximately 760 millimeters—30 inches—high.) However, when he began attacking the problem of incandescent lighting in earnest during late summer 1878, he relied not on a vacuum but on the use of relatively inert platinum, which needed protection from melting rather than oxidation. Believing that an electromagnetic feedback device would provide adequate protection against melting, he hoped to avoid the necessity of a permanent high vacuum for each light.

The January 1879 experiments, however, demonstrated that a high vacuum could not be avoided, and Edison sent out a series of telegrams in search of one of the relatively new Sprengel mercury pumps— pumps he later said he had known about since 1875 but obviously had not felt worth obtaining.[4,5] The search was unsuccessful, and Edison was limited during the next two months to his old mechanical pump. Why he did not show more concern for this failure is unclear. Also unclear is exactly when the new vacuum era began. Surviving correspondence indicates that he first contacted the New York glassblowing firm of Reinmann & Baetz in March

of 1879, and that they delivered the first (Geissler) mercury pump on March 26.[6] Another Geissler pump was obtained about the same time, from Princeton, through Professor C. F. Brackett, Francis Upton's former teacher. When it arrived is uncertain, but both Upton and Edison later recalled receiving it after the one from Reinmann & Baetz.[7] In any case, it apparently was not in working condition and had to be repaired.

After Reinmann & Baetz delivered the first Geissler pump to Menlo Park, Albert Reinmann himself was to follow to help with the installation, which he described as a bit tricky. As it turned out, it was William Baetz who came to Menlo Park and spent considerable time helping to develop vacuum apparatus. Starting at the end of March or early in April, he came over from New York as often as three days a week to repair existing pumps or create new ones. Between then and mid-August he assembled at least eight different arrangements of Geissler pumps, Sprengel pumps, and combinations of the two.[8] The last were produced after Edison saw an article by William de la Rue and Hugo Muller that described not only a combination of two pumps but also a McLeod vacuum gauge, which Edison added to his arrangement later in the year.[9]

A few words might be included here about these two basic forms of mercury pumps. Geissler's pump was essentially an improvement of the enlarged Torricellian vacuum device developed by the Accademia del Cimento in Italy in the seventeenth century. It consisted of a vertical tube full of mercury, just short of the barometric height of thirty inches, connected at the top to a two-way valve or stopcock (one way leading to the chamber being evacuated, the other to the outside air), and at the bottom to a flexible hose (or equivalent arrangement) leading to a container of mercury. Lowering or raising the container lowered or raised the mercury column. When the mercury column was lowered, the stopcock was opened to the chamber, allowing air from the chamber to expand into the space vacated by the mercury. When the mercury column was raised, the stopcock was turned to allow this air to escape to the atmosphere. The Sprengel pump used a long vertical tube split into two tubes near the top, one leading to the chamber to be evacuated and the other to a vessel of mercury. By proper arrangements, mercury was dropped past the opening to the chamber, pushing air trapped ahead of it in the lower part of the tube into a trough at the bottom which captured the mercury and allowed the air to escape to the atmosphere. The Geissler arrangement was quicker, but the Sprengel system could achieve a higher vacuum, hence the advantage of combining the two.

In any event, working with mercury and glass was tedious and frustrating, and the need for a glassblower virtually constant. Edison tried to persuade Baetz to move to Menlo Park but without success. So he placed an ad in the newspaper, which resulted in Ludwig Boehm's joining the select group of Edison pioneers. Boehm had worked with Geissler in Germany and then emigrated to the United States in 1878 at age nineteen.[10] He brought to the New Jersey rural laboratory a much needed skill (earlier in the year Edison had unsuccessfully tried to persuade Upton to learn the techniques of glassblowing), not just for the pumps but for producing a variety of glassware including experimental light bulbs.[11] He stayed a little over a year and was undoubtedly a critical factor in Edison's success.

New efforts in August 1879 produced a pump capable of reducing the pressure in a bulb to 0.00001 atmosphere. Although Edison later called this "the first pump by which a partially satisfactory vacuum was obtained,"[12] he also testified that, while Boehm executed first class workmanship, his pumps were no more efficient than preceding ones.[13] To explain these contradictory statements we can appeal to Edison's frequently selective memory, which perhaps was influenced in the second instance by the fact that he was giving testimony in a patent dispute and was attempting to discredit Boehm's originality. In any event, while these pumps may have produced "a partially satisfactory vacuum," months of additional research were required to develop the improved and much simplified pump used to evacuate bulbs in commercial production. Pumps continued to cause difficulties— breaking, sticking, or otherwise failing to function—but Boehm's presence assured that experimentation could continue.

It is clear that by mid-August, with a good four months of experience behind them, they had solved the basic problem. By the end of the year a bulb could be evacuated to an acceptable level—a millionth (0.000001) of an atmosphere[14] —in twenty minutes. This piece of experimental apparatus was therefore ready for the final assault.

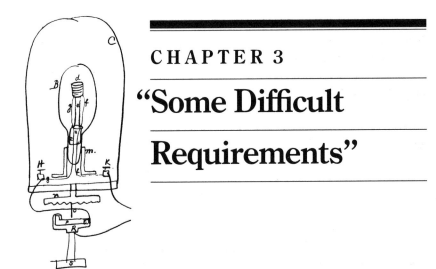

CHAPTER 3

"Some Difficult
Requirements"

Despite the complexity that continued to characterize his efforts, by the spring of 1879 Edison believed that he had finally solved the key technical problems of the platinum lamp. To be sure, the variety of regulators he still employed was a source of trouble and required further work. But the progress he had made in improving the behavior of platinum by vacuum treatment led him to announce that the lamp was essentially perfected and only the remaining elements of the system held up introduction of the incandescent light. This was made public in typically flamboyant Edison fashion, as described in the *New York Herald* for March 27, 1879:

> The first practical illustration of Edison's light as a system has just been given. For the past two nights his entire laboratory and machine shop have been lighted up with the new light, and the result has been eminently satisfactory. . . . Only two things, Edison says, are now necessary before the light can be given to the public. The first is the standard lamp to be used and the second a better generator than the one now in operation. Neither of these requirements is regarded by him as difficult of attainment.

The importance of regarding "Edison's light as a system" was widely apparent—to his backers, his co-workers, and the public at large. That Edison was not simply devising a workable lamp, but building a complex system of lamps, generators, and transmission and control devices, was obvious to all. Too much credit has perhaps been given him for his "systems approach" by twentieth-century observers. It is true that a number of Edison's rivals seem at first glance to have been insensitive to the systems demands of a practical electric light, but there is also little question that his first months of work were hardly more sophisticated than theirs. Edison, like other inventors, saw the real challenge of electric lighting as the invention of a workable lighting element. It was Edison's overwhelming confidence in his ability to meet this challenge that led him to devote more attention to system components.

The gas system provided an obvious model for electric lighting. The complexity implied by this model was readily apparent to nineteenth-century observers. This is best illustrated by a *New York Herald* reporter writing on December 11, 1878:

> The various parts of the system of his electric lighting are probably as numerous and require as many patents for complete protection as did the system of lighting by gas, with its purifiers, gasometers, retorts and the hundred other appliances all going to make up the entire plan. Among the appliances of the electric light which will have to be served before the light as an entirety can be explained are the improved dynamo machines, the regulators, condensers, switches and coolers, besides the different portions of the light proper and the various forms of conductors and lamps to meet the diversity in the wants of the consumers. When all these are completed—and not a day passes without a marked advance toward their completion—the electric light of the wizard of Menlo Park will be ready for inspection, criticism and use, but not before.

To succeed in competition with gas, electric lighting would have to provide comparable service at a competitive price. Throughout his work on the light, Edison kept the competition in mind, gathering data on the sizes and costs of various gas systems, calculating the light output of the typical gas jet, and comparing the economics of gas and electric-arc systems. The gas system also suggested components, such as meters, necessary for any successful lighting system. As early as November 1878 devices were being sketched for measuring electricity consumed by users of the light.[1] A December 1878 *Herald* article quoted Edison to the effect that the successful invention of a meter at Menlo Park was "one of the details that hadn't been accomplished." This appears to have been typical Edison hyperbole, since the Menlo Park records show little progress toward a useful meter before mid-December. On the 15th, Edison made a crude sketch of a meter, explaining that "I propose to shunt a small quantity of the current through a decomposing cell of Ag [silver] or Cu [copper] and weight [sic] the deposit every month to determine the current consumed."[2] This was indeed the form of meter Edison eventually used—a design both simple in its construction and accurate in its measurements. It is doubtful that he had determined the final form of his meter as early as December, however, for he was still making sketches several months later.[3] The broad British specification he drafted in the spring of 1879 included the description of a chemical meter, consisting of

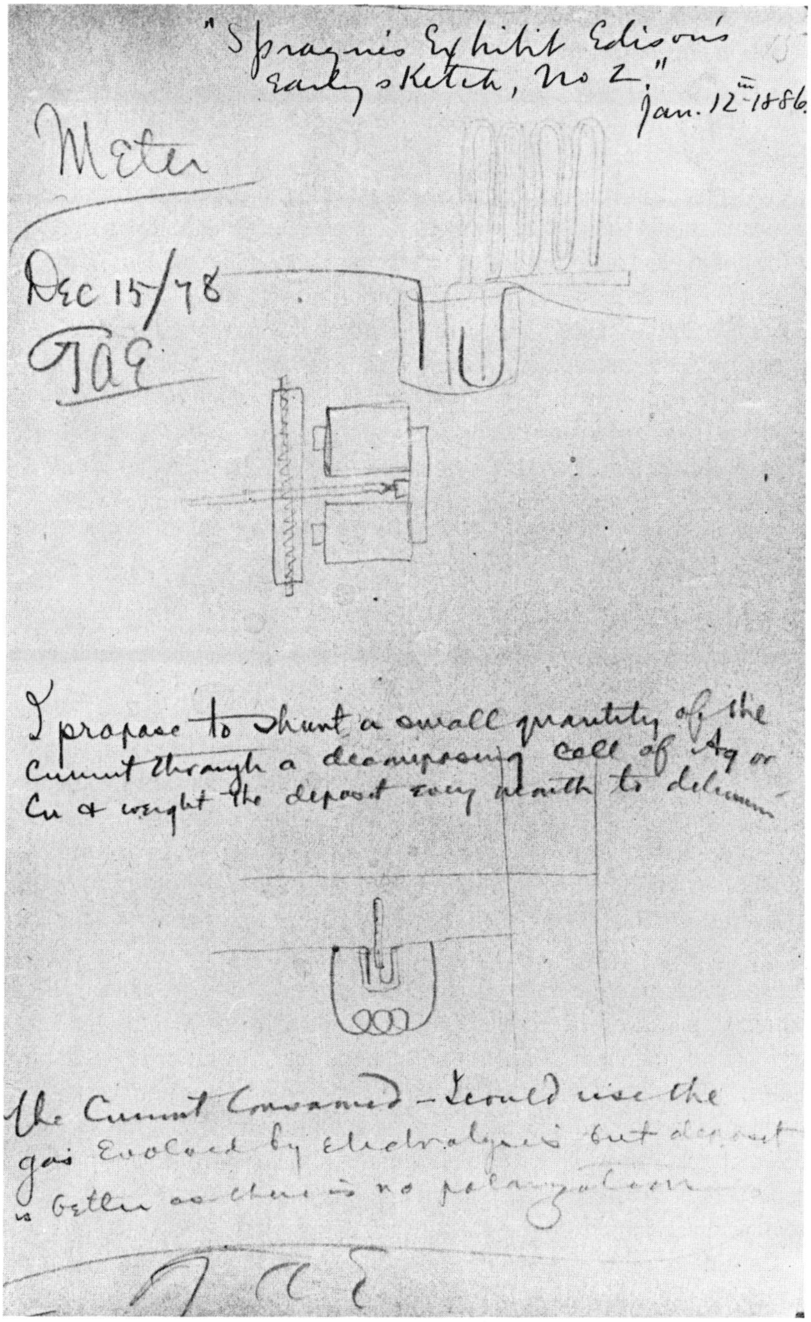

"Sprague's Exhibit Edison's early sketch, No 2."
Jan. 12 1886.

Meter

Dec 15/78
TAE

I propose to shunt a small quantity of the current through a decomposing cell of Ag or Cu & weight the deposit ...

The Current Consumed — I could use the gas Evolved by electrolysis but deposit is better as there is no polarization

Early Chemical Meter, December 15, 1878.
Although Edison's intention to replace gaslighting systems with a comparable electric system suggested the need for a meter, systematic work on meters did not begin until 1880, after the lamp and dynamo had been designed. However, occasional attention was given the problem, as illustrated by this notebook entry describing a meter using an electrochemical decomposing cell, not unlike the type eventually adopted.

copper electrodes in a solution of copper sulphate, to be connected in a building's lighting circuit in much the manner of the first Edison system installations a couple of years later.[4]

Other devices Edison prescribed for his system at this early date included protective mechanisms "to provide against accidental crossing of the conductors leading from the mains" (i.e., short circuits).[5] Once again, the detailed draft of British specifications provides the best description. In the same box with the chemical meter would go a "safety magnet," an electromagnet that was actually a form of circuit breaker. The principle of this novel device is clearly described:

> The object of this magnet is to disconnect the premises from the mains should the leading wires to the lamps be accidentally crossed or get in electrical connection other than through the bobbins of the lamps; the crossing of the wires will tend to draw a powerful current from the mains. The lever B is so adjusted by the retractable spring A by means of the screw C that the current will be insufficient to cause the magnet to attract the lever when all the lamps are in by any great accession of current such as would follow if the leading wires to the lamps became crossed. The magnet would then be powerful enough to attract B . . . [and] the circuit will be opened.[6]

Unlike the chemical meter, the safety magnet was not part of the system Edison eventually introduced commercially. Its place was taken by a simple wire fuse in a wood-block holder.

After late 1878, the team at Menlo Park gave attention to system elements—both specific components and general considerations—with irregular frequency. Progress toward the design of an incandescent lighting system was far from systematic. Laboratory notes and occasional patent applications reveal no pattern in the work on the peripheral elements needed to make the lamp function. Sketches of meter designs are scattered among dynamo experiments. Proposals for rheostats and other control devices pop up in the middle of lamp notes. Sometimes these scattered efforts would bear fruit, appearing as part of a caveat or patent application, or occasionally lying dormant for months until the urgency of imminent lamp demonstrations pressed them into service. Later, after a practical lamp became a reality, various other component problems, previously unanticipated or ignored, would call for an organized attack, but while a workable lamp was still elusive (claims made to reporters notwithstanding), efforts expended on system details were irregular diversions from the main task at hand.

Consideration of the electric light as part of a system like that built around gaslighting raised concern not only about components but also about overall economy. Edison acquired the basic gaslighting journals and gathered statistics on the cost and profitability of various urban gas utilities. He used available figures on total gas consumption to calculate the probable number of lamps in a given locality.[7] He estimated how cheaply gas could be made and distributed and then made numerous reckonings of the anticipated costs of an electrical system. These were used to convince himself that, when all was said and done, he would in-

deed have a competitive system. While there was nothing especially sophisticated about Edison's calculations, the exercise was a remarkable instance of product research. Unlike the other applications of electricity
that had achieved importance—the telegraph, electroplating, the na-
scent telephone—the electric light was not designed to do something no
other technology could do, but rather to compete directly with the well-
established, largely satisfactory gaslight. Hence, the question of whether
the perfected electric light could compete with gas not only technically
but also economically was a very live one. In his public pronouncements,
Edison never tired of repeating the claim that the electric light would be
clearly cheaper than gas. While he unquestionably convinced important
listeners—notably the gas men who backed the Edison Electric Light
Company—he found it necessary to repeat his calculations many times.
He concentrated on the operating costs of the system and this focused
attention even more sharply on the generator.

A typical set of figures was jotted in a notebook in April 1879:

400,000 lights will require if 100 lights per machine, 4000 machines
costing $350 per machine—1,400,000—requiring 68,000 horse
power in 24 stations, 2833 h power in each, cost of this power for
4 hours $2,720. Total cost plant $6,000,000 interest at 10 p.c.
$600,000 yearly.

Interest:	$ 600,000
Power, etc.	992,800
	$1,592,800

If gas consumers would burn in 4 hours at 4 feet per burner
6,400,000 at 2 per M $130,000 roughly or in year 4,745,000 if at
one per M 2,372,500 at this price our cost $1,592,800—our profit
at $1 per 1000, $779,700.[8]

It should be noted that Edison's "profit" consisted of the full difference
between the price of gas and that of electricity—clearly the consumer
was not thought to have much to say in such matters. So much was still
unknown about the final shape of the Edison lighting system that such
figures were, in truth, a bit fanciful. Nevertheless, they reveal Edison's
consciousness of the economic necessities of his promised invention.
More than ever before, technical triumphs would not be enough; truly
competitive solutions would be required.

In the spring of 1879, while the lamp was still far from satisfactory,
Edison nonetheless turned his attention in earnest to the problem of the
generator. The capacity and efficiency of his power source, after all,
would be the key to the economy of his system. He was certain that,
whatever lamp finally emerged from his experiments, it would be an efficient converter of energy into light. He was equally convinced that the

Calculations Comparing Electric and Gas Lighting, c. March 1879.
Edison's determination to make his system competitive with gaslight led to many calculations of relative costs.

243

H.P.
3000.-

18000 lamps

Daily 10 hours

Enginees	9.00
Stokers	7.00
assistants	8.00.
inspectors	8.00,
Coal 27½ tons.	81.50
Extra Coal 3.tons	9.00,
Oil	3.00
Waste	1.00
Repairs,	10.00,
One day 10 hours	136,50,

¾ of 1 cent for 10 hours .

gas.

18000 burners consume,
900.000 feet in 10 hours
which taking the actual cost
of producing the gas at 90°
per 1000 feet,

gas $810.00

136.00.

Electric,

or as compared with Electric Light

Electric light would produce in economy
gas must be made for 15⅛ cents /
per 1000 feet —

generators he had seen and tested—Wallace's, Siemens's, Gramme's, and others—represented a primitive stage of development. With typical cockiness, he set out more seriously than ever to make a fundamental advance in generator technology. Behind his confidence was the feeling that, after several months of work, he finally did know the characteristics of the machine he needed. When he rejected the Wallace dynamo in the fall of 1878, he had done so more from an impression of its general inefficiency than a knowledge of its specific shortcomings as a suitable power source. Now, by February 1879, Edison had a better sense of the kind of generator he wanted and, in the newly methodical spirit he had already applied to the lamp experiments, he brought the resources of the laboratory to bear on the problem.

Edison, Batchelor, and Upton began in February to run systematic tests on the dynamos they had on hand and on modifications of their own devising. A dynamometer was rigged up to measure the work required to run the generators, and measurements were made of resistances, field and current strengths, and resulting electromotive forces. Attention was focused on the design of three elements: commutators, field magnets, and armature windings. Laboratory notes refer often to commutator modifications, although these were directed more at solving the same practical problems of commutation faced by all direct-current generator designers rather than at a significant departure in concept. (Commutators convert the alternating current naturally produced by rotating generators into a single-direction, i.e., direct current.) Many pages of notebooks were filled with armature winding patterns, and the devising of new patterns almost reached the level of a popular sport at Menlo Park during February and March. In the first trials, the armatures of Siemens or Gramme machines were modified, but soon the entire configuration of the generator, especially the size and shape of the field magnets, became the object of experiment. By mid-March, this work produced the basic design for the dynamo that Edison successfully introduced in his first lighting systems. As in almost all of Edison's work in this period, the development of this design owed little to theoretical understanding and much to the ability of the Menlo Park mechanics to execute model after model, modification after modification, as ideas came popping out.[9]

There was a theoretical foundation for Edison's generator work, but it is only dimly perceptible in the many pages of notes and work orders devoted to the problem. This is less a peculiarity of Edison's method than a consequence of the fact that only a small and unreliable body of theory was then available to any would-be dynamo inventor (a state of affairs that did not begin to change until the work of John Hopkinson in the mid-1880s). Edison, like most others, still relied on the almost half-century old conception of the pioneer investigator Michael Faraday, that of a moving conductor (the armature) cutting magnetic lines of force (created by the field magnet)—the more lines of force crossed in the most direct manner, the more productive the generator. The most successful machines of the 1870s, those of Gramme and Siemens, employed armatures with many windings fitted as closely as possible to the poles of an electromagnet. The basic difference between Gramme and Siemens was the pattern of armature winding. The Gramme winding, derived from the ring winding devised by Antonio Pacinotti in 1860, consisted of a hollow cylinder around which the armature wires were laid

Windings for Dynamo Armature, February 27, 1879.
Beginning in February, Edison, Batchelor, and Upton systematically tested dynamos they had on hand and modifications of their own devising. Attention was focused on three areas: commutators, field magnets, and armature windings.

Feb 27 1879

Good Enough

down side by side. Siemens wrapped the armature wires lengthwise around a drum so that each winding crossed near the center. Edison experimented extensively with both approaches and patented dynamos based on both types, but the considerable number of rough armature drawings in the 1879 February and March notebooks reflect increasing attention to the Siemens drum design. Page after page was filled with possible alternatives for drum winding. The reasons for favoring this approach were not spelled out, but the choice was important, for the drum armature turned out to be the basic pattern for almost all significant subsequent generator development.

The adoption of the drum armature, while important to Edison's success, was hardly a breakthrough. That came when the Menlo Park models began to incorporate changes in the generator field magnets. The "long-legged Mary-Ann" dynamo that was to be the prime power source for Edison's lighting system in its first years was distinguished mostly by

Armature Windings for Edison Dynamo, February 17, 1879.
Many pages of notebooks were filled with armature winding patterns, and concocting new patterns almost reached the level of a popular sport at Menlo Park during February and March. Many were tried out in modifications of Siemens and Gramme machines but experiments increasingly concentrated on generators of Edison's own design.

the two large iron poles that inspired its nickname. The source for the idea of placing the dynamo's armature between the poles of a powerful, oversized magnet is not known, but the concept cannot have seemed too subtle to Edison's literal mind. The giant magnet of the generator would be a concentrated source of Faraday's lines of magnetic force, and the turning of a low-resistance armature in such a powerful field would naturally, so it may have been reasoned, be an efficient source of power. The Menlo Park notebooks do not give many clues to the reasoning behind the design, but they do document its emergence and the gradual realization that it marked the end of the search for the right generator. Thus, on March 13, 1879, Charles Batchelor put instructions in the machine shop order book for a generator model: "58. Alter small magneto machine by taking the shell [armature] and putting it in between two powerful poles. So: . . ." A sketch specified that the cores were to be twenty-four inches long and four inches thick.[10] Order 58 was followed by two orders for even larger magnets, one of them specifying "Take the bar of iron that comes tomorrow 13 foot long and cut it in two and make a magnet of it."[11] John Kruesi noted under Order 58 that he finished it on April 12. The results were apparently so interesting that on that date Batchelor put in order 127 modifying what he called the "small Edison Faradic."[12] A week later the design was referred to as "our Dynamo."[13] On May 1st, yet another version of "58" was ordered (Order 150), with poles three feet long and six inches in diameter.[14] This was clearly the emerging design, even though other concepts were not yet entirely abandoned.

For all the simplistic appearance of Edison's development of the long-legged Mary-Ann generator, a comprehension of the principles underlying its successful performance came fairly rapidly. This is revealed in the long and complex British patent specification prepared later in the spring of 1879. Here Edison told not only the details of his design, but the justification for it:

> I have ascertained by means of the Dynamometer for measuring energy applied in foot pounds and by the electrodynamometer for energy obtained in the electric form reduced to foot pounds that the practice of making the leading wires and lamps composing the extraneous resistance equal to the internal resistance of the machine, although the most effective for obtaining the maximum current, but is by no means so economical as when the extraneous resistance is many times the internal resistance of the machine, especially with machines which have a constant field magnet.[15]

The moment when Edison made this important distinction between the capacity of a generator to produce maximum current (which required equal internal and external resistances) and the capacity of a generator to maximize total power output (which called for considerably reduced internal resistance compared to external load) is not clearly marked in the laboratory record, but it was a great step forward in generator design. In practice, this principle was applied by reducing the amount of wire in the drum armature. Edison's significant innovations in the construction of field magnets were also covered by the draft British application, claiming "that great economy is obtained by increasing the electromotive force of the current by the use of powerful field magnets in lieu

54 Make 12 electric [lamps] [as] shown in Book No 4
page 195

March 10th 1879 J.K.
" 17 " J.K.

(55) Make one Magnet (permanent)
× made of 2in round iron
×× made of 1in Square —
the tops made of 2inch steel
and screwed into the top of
core the surfaces on top of
cores and bottom and also the
bottom of steel pieces must be
ground true so as to fit
good. The steel pieces
must be hardened but
not magnetized but left
for me to do that
Batchelor Mch 10th 1879

Steel Steel

2in

××

56. 12 W. Electric Manf Co   200
Carbon Button
(250) Shipped March 12th 1879 J Kruesi
57 Mar 18. W. E Mfg Co   300
Carbon Button

58 Alter small magnet machine by taking the
shell and putting it between two powerful poles.
So —
Cores 24inch
Cores 4   Put the machine on the
wooden base used by big induction
magnets before & fasten down by Z
See book 46. 100 page
Completed April 10th by   March 15th 1879 Batchelor

Design of Large Field Magnets, March 10, 1879.

The development of Edison's bipolar dynamo owed little to theoretical understanding and much to the ability of Menlo Park mechanics to execute modification after modification, as is evident from this book of orders for the machine shop. However, an underlying theoretical foundation can be discerned in the design of the large field magnets. Edison, like most others, relies on Faraday's half-century-old conception of cutting magnetic lines of force with a conductor; the more lines of force crossed in the most direct manner, the more productive the generator.

Edison Dynamo from British Patent, June 17, 1879.

In his British patent Edison stressed the importance of low internal resistance compared with external load to maximize total power output. To accomplish this, he reduced the amount of wire in the drum armature.

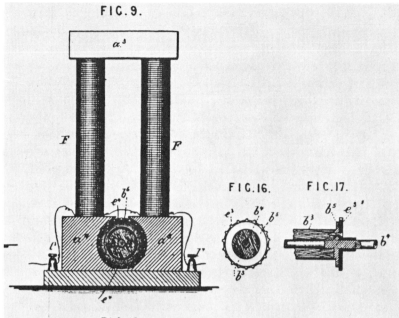

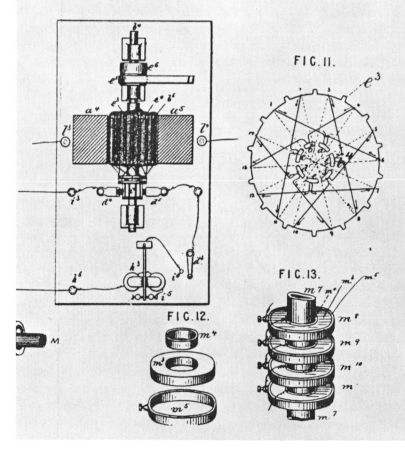

of high speed of the induction cylinders or rings [armatures], such magnets being kept up by extraneous energy."[16] Finally, he also asked patent protection for his use of magnets "of great mass," but without explaining their benefit.

The recognition of the success of the new generator design was revealed in other ways as well. In a letter to his father on April 13, 1879, Francis Upton spoke of frustration with the lamp, but "during the past week we have tried a new magneto machine and found that it was a great success."[17] And at the end of April, Edison spoke in his customarily unrestrained way to a reporter from the *New York World,* boasting "I have had my generator constructed and I tried it for the first time last night. It developed so much power that the coil on the bobbin was torn to pieces and I had to stop."[18] He went on to claim that he would shortly demonstrate the superiority of his machine by matching its performance with that of Gramme's and Wallace's machines, using arc lights for the load. It is a little surprising that Edison would propose such a test, given the special suitability of his dynamo for his proposed high-resistance incandescent system. Indeed, Edison spells out in the same interview the importance of the high-resistance approach to his system: "The resistance of my lamp is as 192 against 1 to the resistance of the carbon [arc] lamp. . . . The point is that the more resistance your lamp offers to the passage of the current, the more light you can obtain with a given current." He then admits that he had not always considered his task as system-building:

> When I first started out on this thing, I took into consideration only the lamp, but I soon became convinced that it was necessary to have a more powerful generator and feasible plan of sub-dividing the light. The generator was the last fact accomplished, and you will soon see for yourself how it works.[19]

At the same time that Edison was announcing the success of his generator and the importance of high-resistance lamps, the U.S. Patent Office granted his patents for the regulator lamps he had devised months earlier.[20] Newspapers duly reported the details—the *New York Herald* filled three columns (April 25, 1879)—which were assumed to represent the state of the art. In its editorial columns, the *Herald* said that "no recent addition to scientific knowledge is more important." The platinum lamps, with their complex regulating mechanisms and fragile strips or spirals of precious metal, were seen to present immediate threats to the gas industry: "It begins to look," the *Herald* continued, "as if the vast capital of the gas companies of all the cities of the world is to be annihilated by the new invention." When a *Herald* reporter hurried out to Menlo Park the next day, Edison said nothing to disabuse the industrious newsman of his incautious optimism. Consequently, the *Herald* editorialized further: "The success of the Edison light opens a new and vast field for enterprise . . . and it will probably rank as the most important scientific discovery of the century."[21] If in the coming months the Wizard of Menlo Park was occasionally looked upon as a humbug, it was not without some justice.

If this judgment seems a bit harsh, the evidence is clear that in the spring of 1879 the search for a workable lamp was entering its most frustrating stage. After months, Edison felt he knew what he was looking for: a light of 8–20 candlepower with a sufficiently high resistance (several hundred ohms, if possible) to be made stable by enclosure in a good vacuum, and with a burner of a heat-resistant material such as platinum. The basic picture of the lamp had not really changed much in Edison's mind since the prototypes of September and October 1878, even if crucial technical details (such as the vacuum and the resistance) had been added. This is perhaps the reason that Edison did little to dampen the *Herald* reporter's enthusiasm for the early regulator lamps when the patents finally appeared in April. He reminded the newspaperman only that the applications had been filed months before and remarked that "we have done some work on the subject since then."[22] In six weeks, he promised, the lamps would finally be ready for public inspection. However, when summer began six weeks later, everything had come to a standstill, the lamp still frustratingly elusive and the pressure of other commitments pushing the electric light aside.

Work on the lamp in the spring of 1879 may be divided into four areas of concern: regulator, material, vacuum, and insulation. The regulator was still looked upon as an inevitable feature of the electric light. Edison simply did not consider making a lamp without a thermomechanical or electromechanical feedback device for temporarily cutting a lamp out of the circuit when its temperature rose close to the fusing point of the incandescing element. The early regulator concepts, both pneumatic and thermostatic, were still the ones he relied on. Because the fundamental problems were thought to lie elsewhere, regulator design was not the subject of steady experimentation; instead, a flurry of ideas would emerge at erratic intervals, often expressed only in sketches of varying degrees of completeness. One such flurry occurred in early April, and rough notebook sketches were sent to Kruesi with orders for models.[23] There is no indication how these ideas fared in the testing, nor is it clear what new departures they represented. It was several months before the record showed another regulator design, and the problem was never again the subject of intensive work.

The focus of lamp efforts continued to be on the "burner"—how to prevent unthinkably high temperatures from destroying a material through melting, volatilization, burning, or cracking. The first attack on

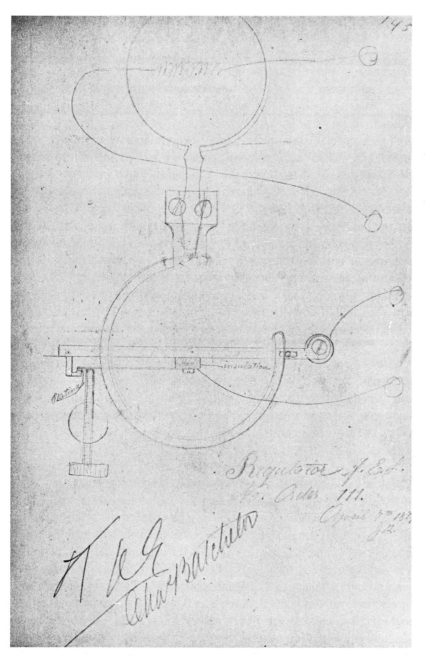

**Regulator Design,
April 7, 1879.**
Because the fundamental problems were thought to lie elsewhere, regulator design received only sporadic attention, and Edison continued to rely on basic pneumatic and thermostatic mechanisms dating from the fall of 1878. A flurry of ideas would emerge at intervals, often expressed only in sketches of varying roughness for use in constructing models. One such flurry occurred in April 1879.

this problem was to eliminate or at least minimize it by finding the most durable available substance. The long list of previous winter experiments on various metals, common and rare, had been part of this effort, but failed to turn up anything better than platinum or platinum-iridium. In mid-April, spirals made of pure iron and of aluminum were put into lamps, but with such unimpressive results that the tests were not repeated.[24] That spring, therefore, the workers at Menlo Park felt there was little more to be gained in searching for alternative substances. As far as Edison was concerned, the electric light would be a platinum light, and time would be better spent securing good sources of the precious metal than in trying to find something cheaper.

Since platinum was deemed the best available material for a lamp, research was to be directed toward lengthening the life of incandescent platinum and making sure it had the required properties, in particular, a high resistance. Vacuum and insulation experiments were aimed at these goals. The first vacuum experiments in February had convinced Edison that the best possible vacuum was indispensable for a long-lived lamp. The Menlo Park lab was equipped with the best pumps available and, in late March, received a new Geissler mercury pump from the New York instrument house of Reinmann & Baetz.[25] Even with these instruments, achieving a good and persisting vacuum in the glass lamp containers was still difficult and time-consuming. It must have been an exhausted Edison who jotted down in a notebook, "We was all night bringing up 12 lamps in vacuum. Worked all day Sunday, all night Sunday night, all day Monday."[26] A month later, Edison wrote in a letter that "my time is wholly taken up from 16 to 18 hours a day with vacuum pumps and time experiments."[27] Others at Menlo Park also found the work painfully slow, Upton writing home that "It is hard to make the lamps we want though we can make very good ones. We have placed our hopes very high for we can make lamps with a few hours of life which are extremely economical and there seems to be no reason why we cannot make them last many hours in time."[28]

By spring 1879, the Menlo Park workers were concerned with more than simply making pieces of platinum last longer in a lamp. The requirement of very high resistance for use in a practical circuit placed limits on the form of the platinum burner. The advantages of high lamp resistance in a parallel circuit had been recognized during the winter experiments, but implications for the lamp's physical design had not actually been worked out. Platinum is not highly conductive. Its resistance is about seven times that of copper for a wire of given length and cross section, and a 20% iridium alloy of platinum has almost twenty times the resistance of copper.[29] Nonetheless, the very high resistances Edison sought required a considerable length of platinum wire in each lamp. In attempting to calculate his requirements, Edison was entering a still murky area. For example, notebook figures probably made sometime in April begin with a measured resistance of 2 ohms for a 16-inch-long piece of 0.009-inch-diameter wire at room temperature (62°F). Edison estimated that the resistance of the same wire with half the diameter (0.0045 inch) would have four times this resistance; one with a quarter the diameter (0.00225 inch), sixteen times, and so forth. Thus, a 16-inch-long platinum wire with a 0.001125-inch diameter would have a resistance of 640 ohms at incandescence; for a 32-inch wire it would be 1,280 ohms, according to Edison's calculations. The conclusion was that "there is no difficulty in making a lamp having 2 or 3000 ohm resistance."[30] For a lamp to have an appropriately high resistance, however, the burner must be an extremely long and thin platinum wire packed into the tightest possible configuration.

This conclusion, firmly based on Edison's and Upton's understanding of the physical laws governing resistance and the energy input and output of the lamp and its circuit, drove the work at Menlo Park down paths no one had ever pursued before, paths that were to prove dead ends but the bold pursuit of which illustrated better than anything else the extraordinary capabilities of the laboratory that Edison had at his disposal. To pack a long length of platinum wire into a tight spiral, it would be necessary to coat the wire with an insulating material, thus allowing successive turns of the spiral to touch one another without shortcircuiting. Much effort was directed toward developing a material that would adhere to the wire, provide good electrical insulation, and withstand the extreme heat of the incandescent platinum. Months before, Edison had concluded that an efficient lamp that gave off relatively little heat for its light would require a very compactly wound wire spiral. It was therefore already standard practice to wind the platinum wire around small spools of compressed lime, which provided a refractory (high-melting-point) support for the spirals. It was widely believed among those at Menlo Park that the discovery of a suitable "pyro-insulator" for the wire spirals was the last remaining step in the creation of a suitable lamp.

The search for an appropriate insulating material was pursued intensively during much of the spring of 1879. In late March, Batchelor considered such substances as "acetate of magnesia, acetate of silica and oxide of cerium." Other candidates included barium nitrate, magnesium nitrate, sodium tungstate, calcium acetate, calcium nitrate, calcium chloride, aluminum nitrate, and zinc acetate.[31] By April, the list of candidates had become extensive and sometimes bizarre: magnesium acetate mixed with rubber, an amalgam of magnesium and mercury, magnesia mixed with gutta percha in chloroform, plaster of paris, silk coated with magnesia, pipe clay mixed with magnesia, and so forth.[32] The coating experiments were in Batchelor's hands, and he was clearly willing to try anything once. Among the substances he proposed in early April were compounds of cerium and zirconium, which were to emerge later as the most promising possibilities of all. The April experiments, however, ended inconclusively and the insulation problem remained a major source of concern.

Frustration with the insulation experiments combined with the press of other business—especially Edison's telephone, which was now being vigorously marketed in England—to push the electric light work aside as

Device for Coating Incandescing Element, February 24, 1879.
To give his lamp sufficiently high resistance, Edison wanted the platinum wire filament to be as long and thin as possible. It also had to be coiled or folded into the tightest possible configuration for space economy and effectiveness as a light source. He devised this instrument for coating the wire with insulation to prevent adjacent turns from touching and causing a short circuit.

spring drew to a close. Even in May, few experiments were performed, and the only significant light-related activity was the commissioning of an international search for sources of platinum. Late that month, Edison informed the Electric Light Company of his plans to send a man through Canada and the western United States to check out unexploited platinum sources. A circular was prepared for wide distribution to western postmasters and Edison responded eagerly to western correspondents who thought they might have what he needed. At the same time, Edison ordered a long list of books on mineral resources in the western United States and Mexico.[33] Finally, Frank McLaughlin, a Newark acquaintance, was sent out, first to Canada and later to the West, to pursue the search. Throughout the summer of 1879 there arrived at Menlo Park ore samples from McLaughlin and from respondents to Edison's circular, and these were all dutifully assayed by John Lawson. Methods were devised for separating platinum from "black sand," and inquiries were made to officials ranging from the Superintendent of the U.S. Mint to the U.S. minister in St. Petersburg, Russia.[34] Since platinum was indeed to play a role in the electric light, though not the one Edison originally saw for it, these efforts were not wasted. Irrespective of their final significance, however, they were the primary light-related activities for most of May, June, and July, as the Menlo Park lab was mobilized to solve urgent technical and production problems with the telephone.

This period marked the most significant lull in electric light work since the challenge had been taken up the previous fall. In May the dynamo received a little attention, and during June and July some time was taken up with dynamometer tests of the latest model "Faradic machine." A few scattered auxiliary devices show up in notebooks—a meter design, notes on conductors and conduits, and a new type of dynamo regulator. The lamp itself was untouched until late July, when laboratory resources were once again brought to bear on a variety of problems, including possible alternatives to platinum, new regulator designs, "pyro-insulation," and perfecting the vacuum pump. A period of dogged experimentation began, characterized not by the testing of novel approaches or devices, but by a persistence in following chosen paths, a persistence remarkable in light of the limited progress that emerged to encourage it. A few individuals were given projects that occupied much of their time for weeks. Charles Batchelor continued his attack on the problem of finding a suitable insulating coating for the long platinum spirals. Francis Upton, assisted by Francis Jehl, devoted himself to testing a simplified lamp design consisting of platinum spirals in high-vacuum bulbs without regulator mechanisms. The record of weeks of other disappointing results reflects, possibly, a new maturity in the Menlo Park approach; no longer was work centered around the testing of flashy new ideas in hopes of quick and easy solutions. On the other hand, there was undoubtedly a depressing air about the work of late summer and early fall, weeks of hard work with barely any sign of progress.

In one area, however, a development would prove crucial to eventual success—the ability to produce better and quicker vacuums. In August, Edison added to the Menlo Park crew a full-time glass blower, Ludwig Boehm, who had apprenticed under Heinrich Geissler himself. For months, experiments had been using both Sprengel and Geissler pumps

Edison's Platinum Search Circular, 1879.

In a New York Herald *interview on July 7, 1879, Edison said he had definitely decided upon platinum burners, was looking for an "unlimited supply of the ore," and was willing to spend $20,000 to find it. He sent this circular to all the major mining areas in the United States.*

FROM THE LABORATORY OF
T. A. EDISON,
MENLO PARK, N. J.
U. S. A

Dear Sir:

Would you be so kind as to inform me if the metal platinum occurs in your neighborhood? This metal as a rule, is found in scales associated with free gold, generally in placers.

If there is any in your vicinity, or if you can gain information from experienced miners as to localities where it can be found and will forward such information to my address I will consider it a special favor, as I shall require large quantities in my new system of Electric lighting.

An early reply to this circular will be greatly appreciated.

Very Truly
Thomas A. Edison

Menlo Park,
N. J.

to evacuate platinum lamps, and a part-time glass blower had been engaged to keep the complicated devices in working order and to make modifications as required. The recruitment of Boehm signified the depth of Edison's belief that a superior vacuum was indispensable to a practical lamp. Both Batchelor and Upton tried their hand at improving the pump design, seeking not only better vacuums but also faster evacuation, since the slowness of the simple mercury pumps put severe limits on the pace of experimentation.[35] After several weeks of work, an intricate, combined Sprengel-Geissler pump was constructed, giving the Menlo Park lab what was perhaps the best vacuum pump then in existence.[36] The new device was very difficult to keep in working order, and its use required careful attention to the sequence of operations (as well as a

The Wizard's Platinum Search, July 9, 1879.
This cartoon from the New York Daily Graphic *depicts the Wizard hunting for platinum. (Courtesy Smithsonian Institution)*

**Early Junction Box Design,
May 20, 1879.**

Work on the electric light slowed perceptibly during late spring and summer. This sketch by Edison is one of the few he made during that period. It shows the design for a Patent Office model of the junction box to be used in his proposed feeder-and-main system of electrical distribution. The junction box later used in the Pearl Street district was very similar to his early drawing.

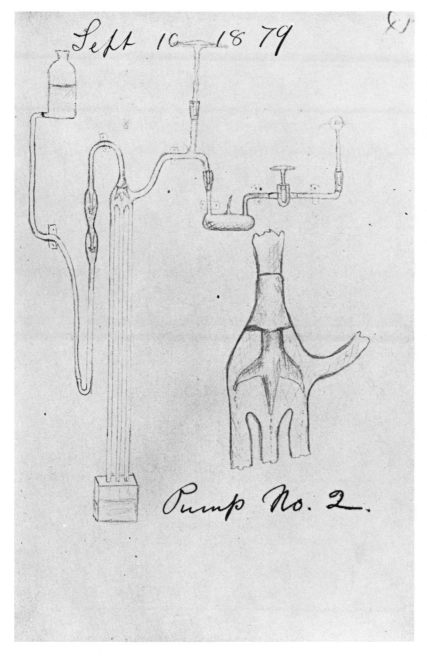

Sept 10 1879

Pump No. 2.

**Experimental Pump,
September 10, 1879.**
*Edison's glassblower, Ludwig Boehm,
drew this pump in his notebook. The
addition of Boehm to the staff, in August 1879, signified the depth of
Edison's belief that a superior vacuum
pump was indispensable to a practical lamp. Although little progress
was made on the lamp that summer,
Boehm helped the laboratory to develop the capacity to produce better
and better vacuums, ultimately crucial to the successful design of the
carbon filament lamp in October.*

good deal of labor lifting mercury, as Francis Jehl vividly recalled decades
later), but the result was an unquestionable success. Edison's conception
of the role of the vacuum in his lamp at this stage is not wholly clear.
There is no evidence that the platinum lamps were troubled by oxidation,
although possibly some of the insulating coatings reacted with hot gases.

Combination Sprengel-Geissler Pump, October 3, 1879.
The Menlo Park laboratory's combined Sprengel-Geissler vacuum pumps were probably the best then in existence.

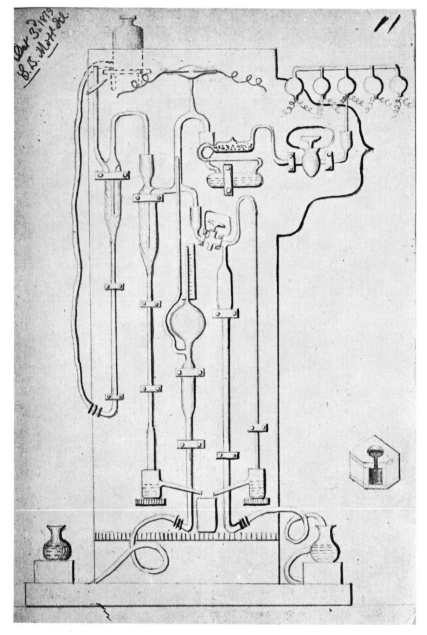

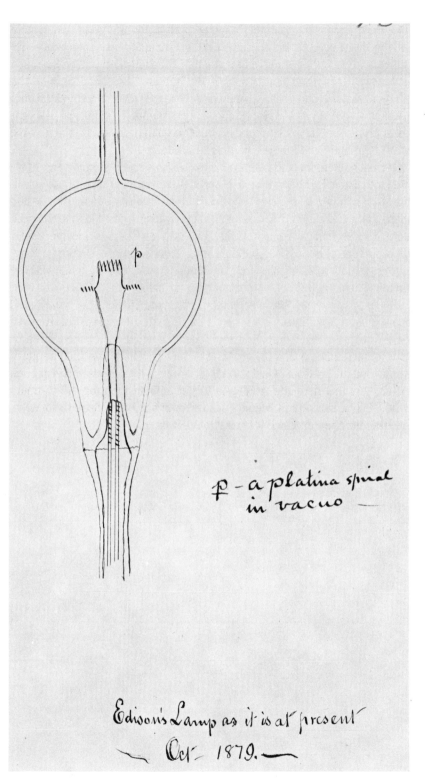

p — a platina spiral
in vacuo —

Edison's Lamp as it is at present
— Oct 1879. —

Platinum Lamp, October 1879.
*This Edison lamp is much simpler
and more elegant than a year earlier.
Gone is the complex mechanical regulator that had so long seemed indispensable. The result looks much like
the modern light bulb. Unfortunately, it still didn't work.*

The most promising coatings, however, were not affected (being stable oxides themselves), and the interest in a vacuum was, in any case, independent of the coating experiments. Ridding platinum of occluded gases was the original function of the vacuum experiments, but the time-consuming tests that Upton and Jehl ran on platinum spiral vacuum bulbs during August and September suggest a purely empirical basis for continuing efforts. Whatever the reason, platinum spiral lamps gave more light and lasted longer in a high vacuum.[37]

By early October the Menlo Park lamp was much simpler and more elegant than it had been a year before. It consisted of a platinum spiral, containing only a few inches of wire, mounted in the middle of a sealed glass globe, exhausted to the best vacuum ever known. Gone was the complex mechanical regulator that had so long seemed indispensable. Gone, even, was the lime bobbin thought necessary to support a long, tightly wound wire spiral. What remained looked, in fact, much like the modern light bulb.[38] Unfortunately, it still didn't work. Once or twice bulbs of this type successfully yielded eight candlepower, but none lasted more than a few hours or tolerated more than the minimal current required for incandescence. And their resistance did not exceed three or four ohms, making a mockery of Edison's carefully specified requirements for his system. On October 8 Upton wrote in his experimental notes, "At this date the trouble is to get an insulation for the platinum wire."[39] The workers at Menlo Park were, in fact, very close to success, but the direction in which it lay still eluded them.

Carbon and the Incandescent Lamp

Because of its resistance and high-temperature qualities, carbon was a natural choice for use in an incandescent lamp. Precautions had to be taken, of course, to prevent it from oxidizing by enclosing it in either an inert gas or a vacuum. But a majority of the early serious experimenters in incandescent lighting still concentrated on carbon, and it is not surprising that Edison started out the same way.[1]

Evidence for Edison's interest in lighting prior to the fall of 1878 is unfortunately limited largely to court testimony, which is suspect because it may be self-serving. However, the activities described are consistent with other known information, and the testimony was in many cases supported by documents which, if no longer extant, were at the time subjected to court scrutiny. What follows is a reasoned reconstruction of what probably happened.

The record provides no specific motivation for Edison's original interest in the electric light. It is not hard to believe, however, that notice of work being done by others in the early 1870s piqued his curiosity. Additionally, publicity being given to arc lighting—including the problem of its being too bright and hence the need for "subdivision"—may have stimulated him.

Although Edison claimed that he had "experimented more or less since 1864 with the electric light"[2] and with incandescent carbon as early as 1869, the first specific reference to lighting experiments is by Kruesi in a personal memo book (apparently since lost) presented in court. Notations indicated that such work was done in Newark on January 5 and 6, 1877.[3] There is no mention of the incandescing material, but, since vacuum apparatus was apparently used, it was probably carbon. Further details are lacking, and in 1881 neither Kruesi nor Edison could recall anything beyond what was recorded.[4]

Carbon was certainly not unknown in electrical laboratories. At Menlo Park in the summer and fall of 1876 paper and cardboard were being carbonized in bulk quantities, not only for use there but also for the production of wires, resistances, battery electrodes, and other items to be sold to the American Novelty Company, which had recently been founded in New York by Edward Johnson who came into Edison's employ at Menlo Park in 1880.[5] In 1877 more carbon was needed for Edison's telephone transmitters.

Batchelor testified, and Edison confirmed, that in August or September of 1877 he cut strips from one or more carbonized sheets and brought them up to incandescence in a vacuum. As Edison recalled, the carbon oxidized, a result they then tried to prevent by coating it with molten glass. The apparatus used was described and also produced for the court (but, as far as is known, has since disappeared). It was a Gassiot Cascade, a relatively common piece of vacuum demonstration apparatus, purchased some time before and modified to fit their needs. What they had in the end was an arrangement of two brass rods supporting the incandescing material under a glass cover.[6]

Batchelor described arrangements for the experiment in some detail.[7] He added binding posts and clamps to the rods. Edison wanted to use hard carbon for the incandescing element, but it proved impossible to get a piece small enough to fit, so they tried the carbonized paper instead. Still, it was a difficult, tedious task to fit the elements in place. As Batchelor testified:

I did it by unscrewing the ball from the top of the rod and also unscrewing the globe from the holder above the cock; also unscrewing the packing cap. When these are all apart the top rod will drop out and the bottom rod can be left in the part having the cock. The carbon was now screwed to the clamp of the bottom rod whilst lying on the table. The other clamp was then screwed to the other end of the carbon, and all three together lifted and turned, so that the part having the cock would be topmost. The lamp also was turned upside down, and the rods and carbon carefully dropped through it. The top rod was then held until the packing and packing cap were put on, when the whole was screwed together again, and the ball replaced. The binding posts were put on the lamp to hold the connection wires from the battery.

The apparatus was then placed on an air-pump plate, evacuated, and brought to incandescence. This was done at least four times (according to Batchelor), or two or three times (according to Edison). The size of the carbon element was three-quarter to one inch long by one-sixteenth inch wide by 0.007 to 0.008 inch thick (Edison), or one inch long by three-sixteenth inch (or less) wide by 0.008 inch thick (Batchelor). Resistance was not measured.[8] Eight years later, in 1889, Edison recalled that these experiments included one with a loop of carbon. He claimed that they cut a strip of cardboard, bent it, and then carbonized it—which is not in accord with the earlier recollections and appears very much to be retrospective wishful thinking.[9]

Further lighting developments in 1877 were documented by papers presented in evidence. Two, dated November 1 and December 3, indicated the use of silicon, boron, and other substances in place of carbon.[10] Another paper, dated October 5, indicated an early understanding of the value of parallel circuits.[11] As Edison noted in his testimony, they:

tried boron, ruthenium, chromium, and the almost infusible metals for separators in my electric light devices. Boron is very high resistance, and would do if arranged thus (diagrammed in parallel) . . . silicon, on the other hand, is of very low resistance, and would have to be arranged thus (in series).[12]

Edison apparently did not return actively to the subject of lighting until the fall of 1878, when, as he prepared for an intensified effort, the above-mentioned 1877 papers were assembled with others and copied into the newly begun series of notebooks labeled *Experimental Researches*. Interest in carbon was revived, according to Edison, at the same time. This was confirmed by Batchelor, who testified that he coated tissue paper with lampblack and tar, rolled it up into rods, and tested its incandescence by heating it in a vacuum. Slivers of wood and broom corn were also tried.[13]

In a discussion early in 1879 on the need for a high resistance element, according to Batchelor, Edison "remarked how easy it would be to get this resistance if carbon were only stable."[14] Then the subject was dropped.

In available documents, carbon for a lamp is next mentioned in a letter, postmarked July 25, 1879, to Edison from Aaron Solomon in California.[15] Solomon wrote that he had assisted a gentleman in England (presumably Joseph Swan) in an exhibition of electric lights with incandescing elements of carbon, platinum, and something said to be better than platinum. Edison's response indicated interest only in platinum, going on to ask if Solomon knew of any sources for it in California.

A brief account of Swan's experiments appeared in an article by J. S. Proctor in the

Newcastle Chemical Society Journal. It began:

At our meeting in December, 1878, Mr. J. W. Swan exhibited an electric lamp, on the incandescent principle, which had broken down in consequence of the electric force being too great for the cylinder of carbon through which it had to pass. One of the points of interest noted was the appearance of a sooty deposit on the inside of the glass. The flask which contained the carbon pencil and its platinum conductors, having been filled with nitrogen and exhausted with a Sprengel pump, was supposed to convey by chemical means any carbon from the incandescent pencil to the cooler surface in its neighborhood.[16]

The rest of the account concerned Proctor's examination of the deposit.

Edison refers to this article in an undated notebook entry[17] which—from circumstantial evidence—was probably not made before January 1880 (see Bibliographical Note). In the meantime, of course, he had turned irrevocably back to carbon which, with the help of the high-vacuum modified Sprengel pump, provided a splendid solution to his problem.

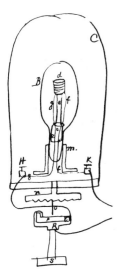

CHAPTER 4

The Triumph
of Carbon

Much has been written about the events at Menlo Park in October of 1879, the most pivotal month in all of Edison's electric light work. Despite the attention given the activity of those autumn weeks in accounts ranging from contemporary newspaper descriptions to the latest Edison biographies, many questions remain about the new turns taken by the research on the electric light. The uncertainty still surrounding the developments of that October is due in no small degree to Edison and his colleagues, who chose to romanticize the activity in the many subsequent years of description and explanation—in large part, no doubt, to conform to popular notions of an inventor's moment of triumph. Unfortunately, biographers and other scholars have also been influenced by such notions, so that most popular accounts weave together indistinguishably the threads of simplistic reminiscences, sensational journalism, romantic suppositions, and incomplete documentary evidence. Because the written record is indeed less than satisfactory, it may not be possible to do much better than make the threads of the oft-told tale of the final triumph of the incandescent light a bit more distinguishable in their varied origins and foundations. Doing so can make clear most of what went on in the laboratory that October.

The first description of the October success was published a mere two months later, when *New York Herald* reporter Marshall Fox, with the cooperation of Edison and the active assistance of Upton, scored the biggest scoop of his career with his full-page story headlined "Edison's Light. The Great Inventor's Triumph In Electric Illumination." The appearance of the story in the Sunday *Herald* of December 21, 1879 was said to have dismayed Edison, who felt it was premature to be put in the position of promising public demonstrations. If so, it was the first time he had shown such reticence to having the press announce his victories, real or imagined. Nonetheless, Edison later said that Fox's account was

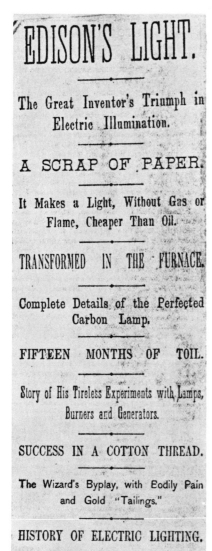

EDISON'S LIGHT.

The Great Inventor's Triumph in Electric Illumination.

A SCRAP OF PAPER.

It Makes a Light, Without Gas or Flame, Cheaper Than Oil.

TRANSFORMED IN THE FURNACE.

Complete Details of the Perfected Carbon Lamp.

FIFTEEN MONTHS OF TOIL.

Story of His Tireless Experiments with Lamps, Burners and Generators.

SUCCESS IN A COTTON THREAD.

The Wizard's Byplay, with Bodily Pain and Gold "Tailings."

HISTORY OF ELECTRIC LIGHTING.

The near approach of the first public exhibition of Edison's long looked for electric light, announced to take place on New Year's Eve at Menlo Park, on which occasion that place will be illuminated with the new light, has revived public interest in the great inventor's work, and throughout the civilized world scientists and people generally are anxiously awaiting the result. From the beginning of his experiments in electric lighting to the present time Mr. Edison has kept his laboratory guardedly closed, and no authoritative account (except that published in the HERALD some months ago relating to his first patent) of any of the important steps of his progress has been made public—a course of procedure the inventor found absolutely necessary for his own protection. The HERALD is now, however, enabled to present to its readers a full and accurate account of his work from its inception to its completion.

the most accurate description published at the time. Since this account had the air of an informed insider's view of the "triumph" and provided an obvious foundation for subsequent stories, it is worthwhile looking at Fox's picture of the state of affairs at Menlo Park in October:

The lamp, after these latter improvements, was in quite a satisfactory condition, and the inventor contemplated with much gratification the near conclusions of his labors. One by one he had overcome the many difficulties that lay in his path. He had brought up platinum as a substance for illumination from a state of comparative worthlessness to one well nigh perfection. He had succeeded, by a curious combination and improvement in air pumps, in obtaining a vacuum of nearly to one millionth of an atmosphere, and he had perfected a generator or electricity producing machine (for all the time he had been working at lamps he was also experimenting in magneto-electric machines) that gave out some ninety percent in electricity of the energy it received from the driving engine. In a word, all the serious obstacles toward the success of incandescent electric lighting, he believed, had melted away, and there remained but a comparative few minor details to be arranged before his laboratory was to be thrown open for public inspection and the light given to the world for better or for worse.

There occurred, however, at this juncture a discovery that materially changed the system and gave a rapid stride toward the perfect electric lamp. Sitting one night in his laboratory reflecting on some of the unfinished details, Edison began abstractedly rolling between his fingers a piece of compressed lampblack until it had become a slender filament. Happening to glance at it the idea occurred to him that it might give good results as a burner if made incandescent. A few minutes later the experiment was tried, and to the inventor's gratification, satisfactory, although not surprising results were obtained. Further experiments were made, with altered forms and composition of the substance, each experiment demonstrating that at last the inventor was upon the right track.

Like all previous experimenters on the incandescent electric light, Edison had tried carbon very early in his work. He claimed to have tested carbonized paper as early as 1877, but found that it burned up almost immediately with even a very small current. Experimenters had met with this result as far back as Humphry Davy in 1802, so Edison wasted little time on carbon and went on in search of a material such as platinum that resisted oxidation even at the high temperatures of incandescence. Just how and why the focus at Menlo Park shifted back to carbon that October is not completely clear, but the shift marked the final step from frustration to success.

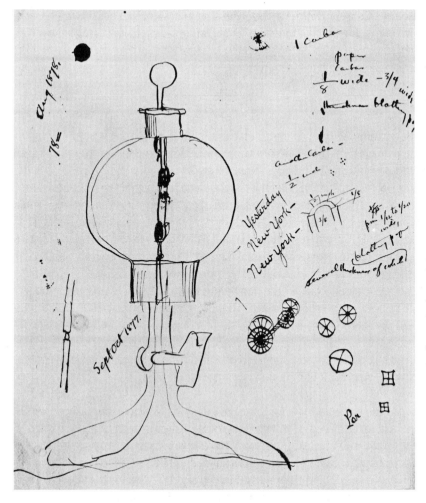

(opposite)
New York *Herald* Lamp Story, December 21, 1879.
Reporter Marshall Fox, with the cooperation of Edison and the active assistance of Upton, scored the biggest scoop of his career with the first published description of the successful carbon lamp of October 1879. Edison was said to have been dismayed by this premature disclosure. If so, it was the first time he had shown such reticence to have the press announce a triumph, real or imagined.

Early Carbon Experiments, 1878.
Edison claimed to have brought strips of paper to incandescence in a vacuum as early as September 1877. A year later, he made this drawing of those experiments.

The renewed attraction of carbon possibly came about somewhat as Fox described it—an unanticipated outcome of the handling of a material very common in the Menlo Park lab those days, thanks to continued commitments involving Edison's carbon telephone transmitter. What Fox left out, and what can be pieced together only imperfectly, is the train of reasoning that led from a crude thin rope of lampblack in the inventor's hand to a carefully shaped incandescing filament in an evacuated glass bulb. To reconstruct this reasoning, it is necessary to bear in mind what Edison truly believed he had accomplished by the fall of 1879, and what he knew he still lacked.

The lamp, as of the beginning of October, was certainly not "in quite a satisfactory condition." Upton, Batchelor, and others were still absorbed in efforts to coat platinum wire "burners" with an insulating compound that could withstand the extreme heat of the lamp and also adhere firmly to the wire. In truth, except for the remarkable improvement in vacuum apparatus, the lamp was little advanced from the preceding spring. Edison was surely aware of this, and this fact alone suggests why he was willing to try such a long shot as his lampblack filament. But failure was not the only spur. After working for more than a year on the light, Edison believed he had learned some very valuable lessons about the end he had in mind. One of these lessons, announced very early in the search, was his "electric light law," which held that the radiating surface of an incandescing element and the ratio of temperature to power input had to be maximized to produce an efficient light. The consequence of trying to maximize both these factors soon led Edison to adopt the coiled spiral as the best shape for his light-producing element. A little later, the conviction that a practical electric lighting system required high-resistance units in a parallel circuit reinforced the concentration on spirals, since this shape maximized the length of thin wire (and hence the resistance) that could be put in a single lamp. Making a spiral from metallic wire presented no problem, but making one from carbon was obviously less straightforward.

On October 7, 1879, the day before Francis Upton wrote in his laboratory notes that finding an insulation for platinum wire was the outstanding problem, Charles Batchelor sketched in another notebook some possible ways of making carbon spirals. The extent to which carbon spirals were considered analogous to the familiar platinum spirals is evident from Batchelor's comments:

> Made a mould for squeezing, put in some Wallace soft carbon, and squeezed it out of a hole .02 [inch] diameter getting it out a yard long if required. Could make more even sticks by rolling on glass plate with piece of very smooth wood. These sticks could be rolled down to .01 and then wound in spirals. We made some and baked them at a red heat for 15 minutes in a closed tube. When taken out they were hard and solid much more so than we expected and not at all altered in shape. A spiral made of *"burnt lampblack"* mixed with a little tar was even better than the Wallace mixture. With a spiral having 5 inches of wire of .01 we can get 100 ohms. We now made a double spiral on brass so as to wind the carbon so similar to some of the first platinum spirals we made.[1]

The two key, interrelated requirements for any possible filament material were high resistance and the capability of being wound into a spiral. Batchelor's notes of October 7 strongly suggested that lampblack could be made to qualify, but the workers at Menlo Park were in no hurry to prove that this was a path to the desired lamp. A few days later, on the 11th, Upton put a carbon stick in a Wheatstone bridge (resistance-measuring circuit) and measured its resistance as it was heated. An 8-inch-long rod, 0.06 inch in diameter, yielded a resistance of only 5.5 ohms which decreased as the rod was heated by the current. This was

**Carbon Spirals,
October 7, 1879.**
Charles Batchelor's sketches of carbon spirals are similar to those of earlier platinum wire spirals, reflecting the assumption that a filament material had to be capable of being wound in a spiral so that enough length could be incorporated to increase its resistance.

obviously disappointing, since all pure metals increase in resistance on heating, the desired effect in a lamp. Upton jotted down Edison's surmise "that pure carbon would increase its resistance on heating. That the cause of ordinary carbon decreasing is that the fine particles make better contact with each other [when heated]."[2] He was sufficiently encouraged to note that he was "trying to mold sticks of .010 in diameter and to make them into spirals." But this was not assigned a major priority— subsequent notes over the next few days made no reference to carbon, involving instead silica insulation for platinum spirals and difficulties encountered with the vacuum pump.

In reconstructing the events of late October, it is important to understand the division of labor then in effect at Menlo Park. While Upton, assisted by Francis Jehl, was devoting all his time to the incandescent light experiments, Edison, Batchelor, and others were deeply involved in a number of projects, some much more pressing than the light. Of particular importance were the demands of Edison's telephone enterprises. In terms of immediate applicability, the most important inventions that had emerged from Menlo Park were Edison's contributions to telephone technology—the carbon-button transmitter and the chalk-drum receiver. Three years after Alexander Graham Bell had introduced his first device, the highest state of the telephonic art in 1879 was a combination of Bell's electromagnetic receiver and Edison's carbon transmitter. The superiority of the carbon transmitter over all rivals was widely recognized, and Menlo Park served as the primary source for the carbon button that was at its heart. In the fall of 1879, not a week passed without requests for hundreds of the compressed carbon buttons, which were produced by a crew of men working in a shed filled with smoking kerosene lamps.[3] Menlo Park always served not only as a research laboratory but also as a small factory, depended upon as a source for the advanced-technology products of the 1870s.

Even more distracting that October were the demands of Edison's other telephone device—the chalk-drum receiver. Exemplifying a phenomenon that was to be common in the high-technology industries of the twentieth century, the chalk-drum device was less a real contribution to telephone technology than a means for avoiding patent conflicts. The Bell ownership of the basic patent on undulatory (wave) transmission put a major obstacle in the way of any rival in the field, thus frustrating exploitation of the Edison transmitter. When the Edison Telephone Company of Great Britain, Ltd., was established in 1878, the narrower British patent laws, which protected devices but not basic principles (such as undulatory transmission), allowed the use of the Edison transmitter but required an alternative to the Bell receiver. The chalk-drum device was Edison's remarkable answer to the problem, for this receiver completely avoided the electromagnetic techniques of Bell. The "electromotograph" telephone relied on the observation that Edison had made years earlier that the friction of an electrode moving over a moist chalk surface varied with the current flowing between the two. The receiver he constructed on this principle required the user to turn a crank connected to the small chalk drum, but the device rewarded the listener's trouble by producing a much louder sound than the Bell telephone. The Edison interests in England took up the new receiver gratefully, and the professional community was hardly less enthusiastic at first, the journal *Engineering* declaring the Edison receiver and transmitter "the most perfect telephonic system that has yet been put forward."[4]

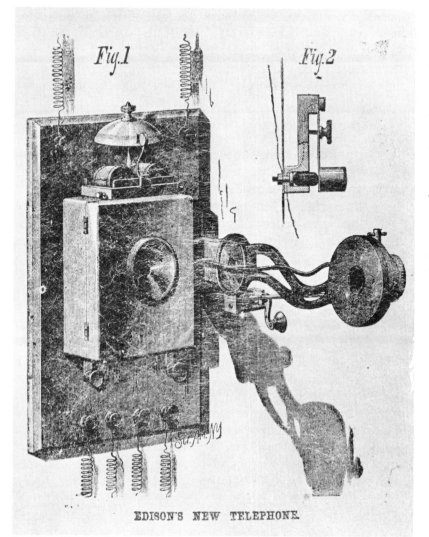

EDISON'S NEW TELEPHONE

Edison's Chalk-Drum Telephone, 1879.
In the summer of 1879, the Menlo Park Laboratory directed much of its attention to the development of the chalk-drum telephone receiver for the English market. This may have indirectly influenced the decision to experiment with carbon in October as the carbon buttons for the transmitters of these telephones were manufactured at the laboratory, and constant handling of lampblack for the buttons may have suggested that it could be rolled into lengths of thin diameter and wound into spirals much like platinum wire.

While the chalk-drum receiver did indeed give the Edison telephone a protected route to European markets, it turned out to be a difficult and balky device in practice. The carbon transmitter was wonderfully simple and reliable, but the new receiver was hard to use and maintain. Customers resisted having to keep turning a crank while carrying on a conversation. Worse, the receiver turned out to be delicate and temperamental, especially since it was difficult to keep the chalk drum properly moist. Dry drums were also a problem in shipping, often arriving in England cracked or broken.[5]

The crew at Menlo Park found themselves occupied with trying to solve the problems of chalk-drum manufacture, shipping, and use. During the early fall of 1879, Edison was devoting much of his own time to these problems, and Batchelor, along with most of the laboratory staff, was having to divide his time between light research and the telephone business.[6] For this reason, laboratory notes reveal very little electric-light activity by Edison for the entire month of October and a two-week lapse between the 7th and the 21st for Batchelor.

On October 19, following several days of difficulties with the vacuum pump, Upton wrote in his laboratory notes that he had returned to carbon: "A stick of carbon brought up in a vacuum to 40 candles (say). Mr. B. trying to make a spiral of carbon. Grease in the pump. Gauge broken. Trying to make carbon spirals." A couple of days later, on the 21st, Upton added to these remarks, "Stick of carbon about .020 and 1/2 inch long gave cold 4 ohms, incandescent 2.3 ohms, very good light. Pt. wires melted."[7] This same day, "Mr. B"—Batchelor—resumed the record of his own carbon-lamp experiments, proceeding toward the actual breakthrough by his customary thorough and careful steps. He first described a series of attempts at making spirals from the tar and lampblack putty. The putty itself was of no use in a light, since it contained a number of compounds that either evaporated or melted at high temperatures. The putty spirals had therefore to be carbonized—baked in an airless container until volatile compounds were driven off and only a skeleton of pure carbon remained. The carbonizing stage, however, was always the stumbling block in the forming of usable tar-lampblack spirals, and Batchelor proceeded to investigate the nature of the problem. When he tried to carbonize pieces of the putty in a glass tube, he was able to observe the hot putty giving off a "yellow oily liquid" which he guessed was "benzole" (i.e., benzine) or a similar substance. This, he surmised, was a cause of the difficulty in successfully carbonizing spirals.[8]

Making spirals was still a key goal to Batchelor. Indeed, his notes at this point dwelt on the problem to such a degree that they hardly betrayed how close he was to a major breakthrough:

Electric Light Carbon Wire—
A spiral wound round a paper cone no matter how thin always breaks, because it contracts so much. If the heating is done slowly this is modified but with the present proportion of tar and lampblack it will always break. . . . One of the great difficulties is to keep the spiral in position whilst you carbonize it. This might be remedied to a great extent by using a hollow sleeve winding the spiral inside with something to hold the ends whilst they are being fastened to the leading wires.[9]

October 21, as far as the laboratory record reveals, came to an end without the dramatic success that subsequent accounts of the electric light's invention attributed to it.

October 22, however, saw something distinctly different. Batchelor's notes of that morning do not suggest great drama, but the activity at Menlo Park took a definite turn, as all the laboratory accounts make clear. In his notebook Batchelor wrote:

Cotton Thread Filament, October 22, 1879.
This notebook entry, which goes on to say that the cotton-thread lamp lasted thirteen and a half hours, may have been the basis for the mythical forty-hour lamp of later recollection. October 21 was commemorated as the date of invention.

Carbon Spirals—9 A.M.

We made some very interesting experiments on straight carbons made from cotton thread so.

We took a piece of 6 cord thread No. 24ˢ which is about 13 thousandths [of an inch] in thickness and after fastening to Pt wires we carbonized it in a closed chamber. We put in a bulb and in vacuo it

Carbonization Experiments, October 27, 1879.

The successful cotton-thread lamp was not treated as a finished invention but as the beginning of a new experimental path. While Upton embarked on a series of measurements to determine the resistance and power requirements of the carbonized-thread lamp, Batchelor spent the last week of October carbonizing a long list of alternative materials.

gave a light equal to about ½ candle 18 cells carbon [battery], it had resistance of 113 ohms at starting & afterward went up to 140— probably due to vibration.[10]

The note does not say how long this feeble light lasted, but on the same day Upton described similar trials with slightly different details:

Carbon spirals and threads. Trying to make a lamp of a carbonized thread. 100 ohms can be made from an inch of .010 inch thread. A thread with 45 ohms resistance when cold was brought up in a high vacuum to 4 candles for two or three hours and then the resistance seemed to concentrate in one spot. Resistance cold 800 ohms.[11]

These observations carry no hint of triumph, no inventor's "eureka!" to set them off, but unquestionably signify the crucial transition of the electric-light search.

Batchelor continued his experiments that day in his usual methodical fashion, constructing a series of lamps with various types of carbon filaments:

    1–of vulcanized fibre
    2–Thread rubbed with tarred lampblack
    3–Soft paper
    4–Fish line
    5–Fine thread plaited together 6 strands
    6–Soft paper saturated with tar
    7–Tar & Lampblack with half its bulk of finely divided lime work[ed] down to .020—straight one ½ inch
    8–200[is] 6 cord 8 strand
    9–20[s] coats 6 cord—no coating of any kind
    10–Cardboard
    11–Cotton soaked in tar (boiling) & put on[12]

Batchelor's notes reveal that he was seeking not only the best material but also the best configuration for a carbon filament; the need was still strongly felt to put relatively long lengths of filament into a small space. The results of his tests confirmed the significance of the behavior of the carbonized thread. Number 2 (thread rubbed with lampblack and tar) "gave an elegant light equal to 22 candles" when powered by an 18-cell battery. Number 3 "came up to 1½ gas jets," but went out due to a short circuit in the lead-in wires. It was, in fact, number 9—simple uncoated cotton thread—that provided the real triumph. In the middle of the night—1:30 by Batchelor's account—the bulb with the simple carbonized thread was put on eighteen cells and kept on. How much light it ultimately yielded is not stated; Batchelor says only that he brought it up

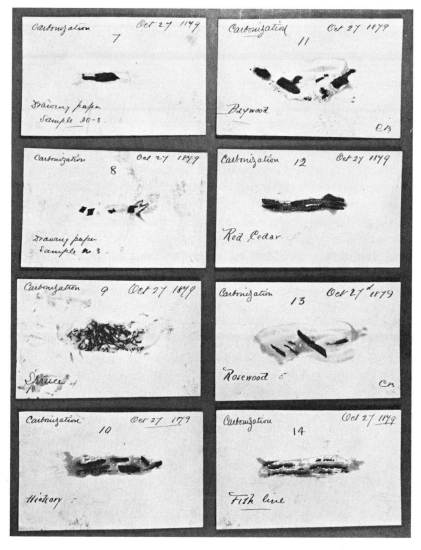

Carbonized Samples, 1879.
Many of the substances carbonized by Batchelor in October were preserved on these small cards.

initially to "½ candle." This, however, would have been too faint to have excited interest, so it is likely that the full eighteen cells made the lamp much brighter. After number 9 had burned for 13½ hours, more cells were added at 3:00 the following afternoon. It continued to burn for an hour longer, yielding a light equivalent to 3 gas jets (at least 30 candles), when the glass bulb overheated and cracked. None of Batchelor's other experiments had results equal to those with this simple carbonized thread.[13]

The Menlo Park response to the success of the cotton-thread lamp, however, was not to treat it as a finished invention, but rather as the beginning of a new experimental path. Edison, of course, appreciated success as quickly as anyone, but the problems posed by the incredibly fragile carbonized thread filament were large indeed, and the stubbornly held belief in the need for a spiral or coiled filament provided little encouragement for the continued use of short, brittle lengths of thread like October 22's number 9. While Upton embarked on measurements to determine the resistance and power requirements of the carbonized-thread lamps,[14] Batchelor spent the last week of October carbonizing a long list of other materials, including celluloid, wood shavings from boxwood, spruce, hickory, baywood, cedar, rosewood, and maple, punk, cork, flax, coconut hair and shell, and a variety of papers.[15] In addition to carbonizing experiments, Batchelor continued to test the series of various lamps he had started on October 22, a series that eventually reached number 260. His tests were not completed until more than two months later.[16] During these two months, Upton and Batchelor, with help from Kruesi, Jehl, and others, devoted an enormous amount of effort to perfecting the carbon lamp and devising means for producing and using it. The transition of the Menlo Park operation from research in quest of a feasible lamp to development of a practical and marketable product was remarkably rapid and smooth, reflecting the basic Edison attitude that inventions were worth something only when they were usable and saleable. This complete changeover, more than anything else, signified the realization of Edison and his men that they finally had an important invention on their hands. Nonetheless, they also saw that there was still some distance to go between what they had accomplished and the system they envisioned as a successful rival to gaslight. November was a time of alternating optimism and frustration, succinctly expressed in Francis Upton's regular letters home. On November 2, he wrote:

> The electric light is coming up. We have had a fine burner made of a piece of carbonized thread which gave a light of two or three gas jets. Mr. Edison now proposes to give an exhibition of some lamps in actual operation. There is some talk if he can show a number of lamps of organizing a large company with three or five millions capital to push the matter through. I have been offered $1,000 for five shares of my stock. . . . Edison says the stock is worth a thousand dollars a share or more, yet he is always sanguine and his valuations are on his hopes more than his realities.[17]

The next week, however, Upton reported, "The Electric Light seems to be a continued trouble for as yet we cannot make what we want and see the untold millions roll upon Menlo Park that my hopes want to see."[18] Finally, a week later, on November 16, Upton's letter home reflected complete confidence in their achievement:

> Just at the present I am very much elated at the prospects of the Electric Light. During the past week Mr. Edison has succeeded in obtaining the first lamp that answers the purpose we have wished. It is cheap—much more so that we even hoped to have."[19]

The light is obtained from a a piece of charred paper which is bent thus

The burner is made from common card board and cut to about the size shown. This is then sealed in a glass bulb and the air exhausted and then a current of electricity passed through it which heats it to brilliant whiteness so that it will give a light equal to that from a good sized gas burner. The making of such

Upton's Description and Drawing of Carbon Filament, November 16, 1879.
Upton, in this letter to his father, described the new carbon lamp and sketched the horseshoe-shaped filament made from carbonized cardboard.

He then went on to describe the cardboard filament lamp in all its elegant simplicity.

Another indicator of the attitude at Menlo Park toward the carbon lamp breakthrough was the patent application for an "electric lamp" filed November 4, 1879 (and eventually granted as U.S. Patent 223,898 on January 27, 1880). Since this was to be the key patent in the Edison system, it merits a careful look. In the beginning, Edison spells out clearly the distinction between his carbon lamp and all others:

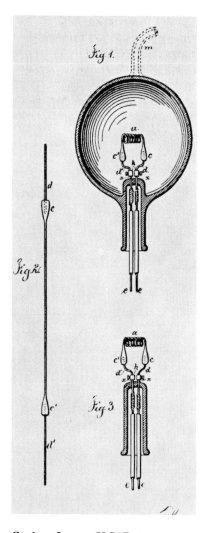

Carbon Lamp, U.S. Patent 223,898.

Edison's key carbon lamp patent showed the filament in the form of a spiral. It is unlikely that a successful lamp was ever made with a spiral since carbonized materials were too brittle to allow such shapes.

The object of this invention is to produce electric lamps giving light by incandescence, which lamps shall have high resistance, so as to allow of the practical subdivision of electric light. The invention consists in a light-giving body of carbon wire or sheets coiled or arranged in such a manner as to offer great resistance to the passage of the electric current, and at the same time present but a slight surface from which radiation can take place. The invention further consists in placing such burner of great resistance in a nearly-perfect vacuum, to prevent oxidation and injury to the conductor by the atmosphere. [20]

After the fashion of such things, the patent then details the making of the carbon filament (this is where the term is introduced) from a variety of materials, conceding the extreme fragility of the finished product. The final form of the filament is invariably specified as a spiral or coil, and the patent drawing depicts such a form. It is highly unlikely, however, that a successful lamp was ever made at the time with a carbon spiral since the carbonized materials were simply too brittle to allow such shapes. The patent also specifies joining the carbon filament to platinum lead-in wires using a lampblack-and-tar putty, rather than clamps. This, too, was apparently wishful thinking, for all lamps made in this period, and for months afterward, required tiny platinum clamps to secure the filament. One other feature for which Edison claimed patent protection, and which was to be surprisingly significant in ensuing years, was the use of a sealed enclosure entirely of glass, dispensing with troublesome metal-glass connections except for the tiny platinum lead-in wires. The basic incandescent lamp patent was only one of literally hundreds that Edison received for his work in electric light and power, but it was unquestionably the most significant, economically and conceptually.

The patent application submitted November 4 did not so much describe what had actually been made at Menlo Park as what Edison and his colleagues knew should be made. The process of learning what *could* be made occupied most of the remainder of 1879, as Edison put the Menlo Park team to work not only to make practical lamps but also to construct enough of a complete lighting system to convince a by now sceptical public and an ever more wary financial community. The attitude that guided the Menlo Park workers is well reflected in another of Upton's letters home:

The Electric Light is slowly advancing from the last big step. We now know we have something and that is what we [did] not know until last week. We can compete with gas in a great many ways now though not as completely as we wish, yet there seems to be nothing to prevent our getting a perfect burner that shall do as well as gas. Time and cost will prove what we have to be good or bad. [21]

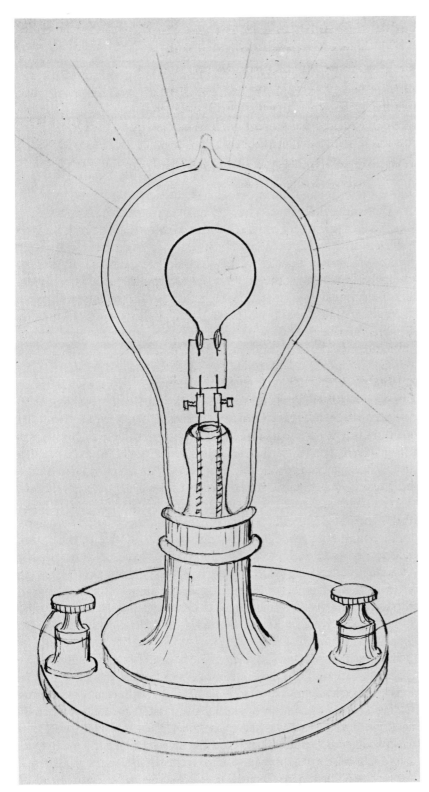

Carbon Filament Lamp, November 1879.
Samuel D. Mott, Edison's patent draftsman, drew this lamp showing the method then used to clamp the horseshoe-shaped filament to the lead-in wires.

There seems to have been little doubt in Edison's mind that what he had was good, and he was determined to show this to the world as quickly as possible. Yet, unlike his habit in the past, he did not talk freely to the New York reporters who had learned that a visit to Menlo Park was almost always good for lively copy. By this time, Edison may have felt that he had been ill-served by his openness with the press, for over the past year the newspapers had been all too ready to report his successful development of the light, only to force Edison to explain to Grosvenor Lowrey and other backers that important problems still remained. This time there would be no premature announcements, no published stories forcing him into making excuses or demonstrating devices not quite ready. When Edison's London representative, Col. George Gouraud, cabled to convey the *London Times'* request that it be given sufficient notice of any Menlo Park exhibition of the light, Edison answered, "Public demonstration takes place during holidays. It is an immense success. Say nothing."[22] As Upton wrote home in early November, Edison quickly decided to put together a full-scale demonstration of the light and its system, but Menlo Park was to keep silent until the demonstration was ready.

No time was lost in assembling a demonstration system. As the work on making the carbon lamp more dependable and durable went ahead with all speed, the details of auxiliary system elements also received attention. On November 4, Edison cabled Norvin Green, president of the Western Union Telegraph Company, to request the services of two linemen to erect wires for a Menlo Park "light exhibition."[23] The Western Union men responded quickly, arranging to come out the next week. Upton, in addition to his continuing lamp tests, was assigned the responsibility of perfecting the chemical meter that Edison envisaged for measuring customer current consumption and put John Lawson, one of Menlo Park's chemists, to work on the problem.[24] Correspondence went out to arrange for the purchase of steam engines for Menlo Park's generators, Edison writing to one firm on November 17, "I expect in 3 weeks from today to have the Park lighted by electricity and the prospect now is that you will soon get an order for an engine or engines."[25] The Menlo Park correspondence of November and December depicts a gathering momentum in activity as Edison mobilized his laboratory to exploit October's breakthrough.

External pressures were added to the laboratory's own enthusiasm as a spur to activity. The most important was from the financiers of the Edison Electric Light Company. In mid-November Edison received from Grosvenor Lowrey a report of the Company's latest meeting at which the need for more money was the major topic of discussion. Lowrey once again soothed the more impatient of the investors, and the meeting resulted in a call for voluntary additional contributions to the Company at the rate of $5 per share from the cash subscribers.[26] Clearly, however, the anxiety of the electric light shareholders had to be addressed as soon as possible. By early December, Edison's secretary, Stockton L. Griffin, wrote to a correspondent:

> Edison is now ready for his exhibition, but will be compelled to wait 3 or 5 weeks for his patents. Upton's house was lighted last night. Edison's will be illuminated tomorrow night for Mr. Fabbri and Mr. Wright.[27]

A few days later, Upton reported to his father on the subsequent demonstration:

> The light is still prosperous; I have had six burners in my house during the past week and illuminated my parlor for the benefit of a party of visitors from New York. The exhibition was a success. Mr. Edison's and my house were the only ones illuminated. There will be a great sensation when the light is made known to the world for it does so much more than anyone expects can be done.[28]

Other pressures for public display of the light mounted as December wore on. The press was still being kept at arm's length, Edison allowing only Marshall Fox of the *New York Herald* access to the lab, and that only on condition that Fox hold his story until Edison was fully ready. Despite this, newspaper reports began appearing early in the month giving word of a breakthrough at Menlo Park and creating wide expectation of an early exhibition. Letters began coming in requesting access to any exhibition as well as further information on the invention itself. For a while Edison tried to keep his correspondents at bay, replying to one on December 9, for example, that the newspapers were wrong in predicting a showing of the light at Christmastime, but an announcement would appear in the *Herald* sometime before February. As the lamp tests continued, however, and other preparations were completed without major difficulty, Edison's caution began to wear off. On December 17, he cabled one of his London agents, "Exhibition ready—capitalists won't allow it until about New Years."[29] Finally, a day or so later, the silence broke and it was announced that a full-scale display of the new light would be open to the public on New Year's Eve.

When or where the promise of a New Year's Eve exhibition was first made public is not known, but when Marshall Fox's famous *Herald* article of December 21 appeared, it was apparently common knowledge. The article itself caught Menlo Park off guard, as Upton reported home:

> Today has been an exciting day since this morning's *Herald* contained an account of the discovery of the lamp and the whole invention. Mr. Edison had allowed a *Herald* reporter to take full notes so

Carbon Filament Lamp, December 1879.

This December drawing by Mott now shows a carbonized paper filament connected to lead-in wires by platinum clamps.

as to prepare his account for the exhibition which was to come off in a few weeks. The reporter was Edison's friend and he thought he could keep a secret. Yet newspaper traditions were too strong and he sold out at a good price I suppose for he had the first full account. Mr. Edison is very much provoked and is working off his surplus energy today.[30]

We can take Upton's word that Edison was annoyed by the unexpected appearance of Fox's report, but it is unlikely that his displeasure lasted long. He, after all, had already told a number of people that he was ready, waiting only for the "capitalists" and the patent attorneys to allow him to go ahead. He recognized the wisdom of having the patent situation settled before explaining too much publicly, and on the 21st he cabled London to alert agents that the *Herald* article had appeared and to urge them to push the European patents through. On the other hand, he seems to have given Fox little difficulty over the article's appearance, for very detailed *Herald* reports continued out of Menlo Park for the next two weeks. Edison was probably pleased to see his backers forced by events to allow the public to see what he had accomplished. The financiers of the Edison Electric Light Company had seen firsthand too many of Edison's breakthroughs of the preceding year fizzle out when put to the test and were understandably nervous, as shown in a letter sent Edison on December 26 by Eggisto Fabbri, the primary representative of the J. P. Morgan interests in the Light Company:

> I suggest to you the wisdom & the *business* necessity of giving the whole system of *indoor* & outdoor lighting *a full test of continuous work* for a week, day & night, *before* inviting the public to come & look for themselves. As long as you are trying private experiments, even before 50 people, a partial failure, a mishap, would amount to nothing, but if you were to express yourself *ready* to give a public demonstration of what you considered a complete success, any disappointment would be extremely damaging and probably more so than may appear to you as a scientific man.[31]

Edison knew, however, that it was too late to call anything off—as Fabbri himself surely must have known. To reassure the investors, Edison gave them a preview at Menlo Park on December 27. Upton reported home that "several million dollars of capital were represented," and the entire show "went off splendidly."[32]

During the last ten days of 1879, the New York newspapers fed their readers regular stories from Menlo Park detailing the comings and goings of important observers and recounting the preparations for the public showing. It is the newspapers, therefore, that offer the most vivid record of the electric light's introduction. On Christmas Eve, it was reported, Edison entertained a number of gas industry representatives at Menlo Park, demonstrating the virtues of his parallel circuit connections for the lights. A matter of particular interest was the number of gaslight equivalents that Edison's system provided per horsepower (delivered by the generator steam engine), since the economy of the electric light was a practical matter still very much open to question. In reply to inquiries and to correct some mistaken newspaper accounts, Edison reported that his system yielded "ten gas jets per horsepower." This led to some confusion, as shown in the interview with Prof. Henry Morton of the Stevens Institute of Technology that appeared first in the *New York Times* and later in other papers. To Morton, Edison's figures confirmed his skepticism about the incandescent light, since, he pointed out, arc-light systems yielded about 100 gas jets per horsepower. The Edison light, Morton concluded, was only one-tenth as efficient as other forms of electric lighting and hence hardly represented the great breakthrough that was claimed. To all this, Edison simply replied that Professor Morton should come out to see for himself, and the professor's refusal brought scoffs from such Edison boosters as the *Herald*, which put Morton down as one who "believes that the world is finished and that there is no room for new inventions."[33]

As December drew to a close, the publicity rose to new heights. The *Herald*, especially, feeling that it had staked out a special claim to cover Edison's success, kept up the drumbeat. On the 28th, the paper reported the success of the demonstration for the Light Company investors, telling how the laboratory was lit much of the evening by forty steadily burning lamps. The next day, it described growing crowds at Menlo Park, wealthy men and ordinary people alike thronging to the New Jersey village to exclaim "Wonderful!" at every turn. On the 30th, further reports came of visitors who went away with "no particle of doubt in anyone's mind that the electric light is a success and a permanent one." And on the final day of preparations for the public exhibition, the *Herald* featured a lengthy description of not only the many visitors to the laboratory, but also the late evening in the lab, after all other outsiders had gone home, when the Menlo Park workers and Edison himself held forth in good humor, apparently content that their task was near completion and that theirs was a job well done.

There is, as might be surmised, no better description of the New Years Eve exhibition than the *Herald's* own, which conveyed the excitement of the occasion and also summed up well the extent to which Edison tried to display not just his light but a glimpse of the whole system he envisaged:

Edison's laboratory was tonight thrown open to the general public for the inspection of his electric light. Extra trains were run from east and west, and notwithstanding the stormy weather, hundreds of persons availed themselves of the privilege. The laboratory was brilliantly illuminated with twenty-five electric lamps, the office and counting room with eight, and twenty others were distributed in the street leading to the depot and in some of the adjoining houses. The entire system was explained in detail by Edison and his assistants, and the light was subjected to a variety of tests. Among others the inventor placed one of the electric lamps in a glass jar filled with water and turned on the current, the little horseshoe filament when thus submerged burned with the same bright steady

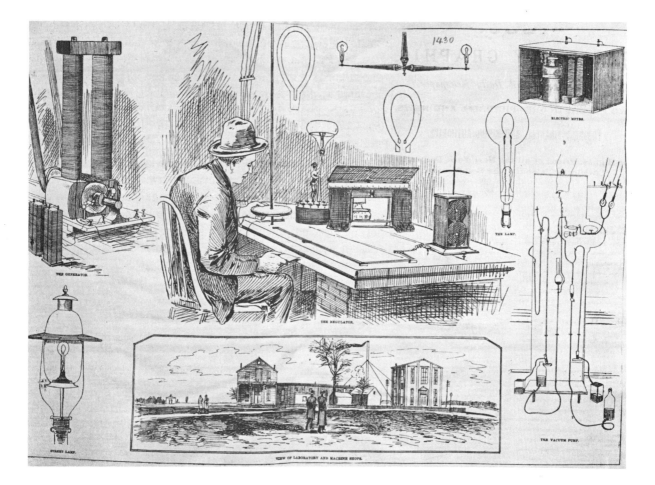

illumination as it did in the air, the water not having the slightest effect upon it. The lamp was kept thus under water for four hours. Another test was turning the electric current on and off on one of the lamps with great rapidity and as many times as it was calculated the light would be turned on and off in actual household illuminations in a period of thirty years, and no perceptible variation either in the brilliancy, steadiness or durability of the lamp occurred. The method of regulating the supply of electricity at the central station was explained in detail, as was also the electric motor; the latter was made to pump water and run a sewing machine with only as much electricity as was necessary to give an illumination of the brilliancy of an ordinary gas jet.[34]

The crowds that thronged into Menlo Park were made up of curiosity seekers and newshounds after the latest sensation. But they also represented something more—a new relationship between advanced tech-

The Electric Light at the Menlo Park Laboratory, December 1879–January 1880.
Drawings of the laboratory in the New York Daily Graphic *showed the lamp as it appeared to the crowds that came to view the new marvel.*

nology and the common man. Edison's electric light was as mystifying and awe-inspiring as any invention of the age. Few things could be imagined that were more marvelous than a piece of charred paper glowing brightly enough in its glass container to light up a room and yet not burning up. But instead of the fear and suspicion such a strange device might have evoked in the unscientific layman of earlier times, Edison's light inspired unalloyed admiration in most of its beholders. The master of Menlo Park had already earned the sobriquet of "wizard," so the average newspaper reader was somewhat prepared for "magic" to emerge from Edison's laboratory, but the wizardry of scientific technology was now a source not of distrust but, rather, of hope. This attitude toward the powers of science and technology is one of the nineteenth century's most important legacies, and no single instance exemplifies it better than the enthusiasm with which the crowds ushered in the new decade at Menlo Park.

Who Invented the Incandescent Lamp?

Edison was by no means the only, or the first, hopeful inventor to try to make an incandescent electric light. The following list, adapted from Arthur A. Bright's *The Electric Lamp Industry,* contains over twenty predecessors or contemporaries. What was different about Edison's lamp that enabled it to outstrip all the others?

Date	Inventor	Nationality	Element	Atmosphere
1838	Jobard	Belgian	carbon	vacuum
1840	Grove	English	platinum	air
1841	De Moleyns	English	carbon	vacuum
1845	Starr	American	platinum	air
			carbon	vacuum
1848	Staite	English	platinum/iridium	air
1849	Petrie	American	carbon	vacuum
1850	Shepard	American	iridium	air
1852	Roberts	English	carbon	vacuum
1856	de Changy	French	platinum	air
			carbon	vacuum
1858	Gardiner & Blossom	American	platinum	vacuum
1859	Farmer	American	platinum	air
1860	Swan	English	carbon	vacuum
1865	Adams	American	carbon	vacuum
1872	Lodyguine	Russian	carbon	vacuum
			carbon	nitrogen
1875	Kosloff	Russian	carbon	nitrogen
1876	Bouliguine	Russian	carbon	vacuum
1878	Fontaine	French	carbon	vacuum
1878	Lane-Fox	English	platinum/iridium	nitrogen
			platinum/iridium	air
			asbestos/carbon	nitrogen
1878	Sawyer	American	carbon	nitrogen
1878	Maxim	American	carbon	hydrocarbon
1878	Farmer	American	carbon	nitrogen
1879	Farmer	American	carbon	vacuum
1879	Swan	English	carbon	vacuum
1879	Edison	American	carbon	vacuum

First, consider the requirements of a successful, individually controlled, moderately bright, incandescent lamp. They were not all obvious at the state of scientific and technical knowledge prevailing in the middle of the nineteenth century. With some advantage from hindsight, the three essential features of a lamp compatible with a practical lighting system are described below.

Incandescent Material

A material was needed that could be heated electrically, without melting or otherwise disintegrating, until it glowed brightly enough to be useful, tolerable by the eyes at close quarters, and comparable to the then familiar gas jet or oil lamp (10–20 candlepower). Most serious investigators worked with carbon, which was readily available, inexpensive, and eminently successful in arc lamps, or metals like platinum, which had a high melting point and was chemically inert. However, inherent problems, many unpredictable at that time and stumbled into the hard way, stood in the path of success.

High Vacuum

Appreciation of the need for a relatively high vacuum (of the order of 0.00001 atmosphere) developed slowly and was in-hibited by the lack of adequate pumps. Certainly, something had to be done to prevent a vulnerable incandescing element like carbon from oxidizing. The simpler expedient of enclosing it with an inert gas proved to cause an unacceptable (even if not recognized) cooling effect. The gas conveyed heat from the lighting element to the enclosing wall, making it more difficult to maintain the element at an efficient incandescing temperature. (Some modern light bulbs contain substantial amounts of gas but have been specially designed to reduce heat losses.) Means for producing a sufficiently high vacuum did not become available until an improved form of the Sprengel mercury pump was introduced in the early 1870s. Even then a hard-won vacuum could be lost when gas trapped in the incandescing material escaped when the lamp was first heated. Pumping had to be continued with the element heated in order to remove occluded gases before the enclosure was sealed. Before 1880 this final step was apparently taken only by Edison and Joseph Swan. Another problem that plagued Edison for a long time was sealing the glass envelope effectively, especially around the lead wires, to ensure retention of the high vacuum. Only Edison seems to have solved such problems satisfactorily. The evidence for Swan and others is not conclusive, but their failure to demonstrate consistently long lamp life suggests that they did not overcome their difficulties.

Electrical Supply System and Lamp Resistance

Finally, there is the question of the electrical supply system and its consequences in terms of desirable lamp resistance, system voltage, and conductor current. Before the 1870s it was natural to assume the use of batteries or magnetos (electromagnetic generators that were relatively inefficient because either the armature or field used permanent magnets). These had technical limitations and high enough costs to make any centralized distribution system economically unfeasible. At best, incandescent lighting would be an oddity for the wealthy or for special applications. But for small systems, with low losses in the lead wires, the resistance of the incandescing element was not a critical factor. Since low resistance was easier to achieve, that is what all the early investigators used. Low resistance lamps inherently imply high current and low voltage.

With the development of practical dynamos (generators in which both armature and field were electromagnets) in the mid-seventies, the situation changed dramatically. It became possible to consider generating substantial quantities of electricity at a central location. However, electric current flowing through long feeder lines creates heat, representing wasted energy and higher cost. To reduce losses, current should be low. This means that the lamp resistance has to be high, since the heat developed in the lamp element (which determines how much light is emitted) is proportional to the resistance times the square of the current. Losses can also be reduced by lowering feeder line resistance, but this means the lines must be short (precluding a central station serving a large area) or fat (trading heat loss for copper cost). High resistance is therefore a desirable lamp characteristic that permits the efficient delivery of electric power from a convenient central station through feeder lines of reasonable dimensions.

High-resistance lamps require relatively high voltages to drive even the relatively low currents through the circuits.

In the 1870s 100 volts was sufficient, although even higher voltages would have been better. At the consuming end, "high" lamp resistance meant about 100 ohms.

Not that there weren't alternative solutions for low-resistance lamps and low voltages. One was isolated generators, each supplying only a single house or factory, so feeder lines could be kept short and losses acceptable. There was a market for isolated systems; Edison himself sold many of them, and Swan was installing them in England as early as 1881. But these used high voltages. No one, including Swan, promoted a low-voltage system, which surely suggests a lack of confidence in available low-resistance lamps. Another alternative was to distribute electricity at 100 volts or more and reduce it to a few volts at the destination. Such reduction became economically feasible in the mid-eighties after the development of a transformer, but transformers

are alternating-current devices and it was a direct-current age. Other ways of manipulating direct-current voltage levels were possible at the time, but none was actively pursued, contributing further evidence for the absence of a practical low-voltage lamp.

Did Edison invent the incandescent electric lamp? He undoubtedly learned something from others, but he stood alone in his appreciation of the essential requirements, set his goals accordingly, overcame many obstacles that stalled his rivals, and developed not only a practical lamp but the associated components, such as improved generators and other hardware, that made a large-scale lighting system possible. And then he built the system.

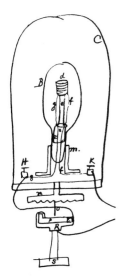

CHAPTER 5

Business and

Science

On New Year's Eve, 1879, Edison put on public display not simply his carbon lamp but the first detailed version of a complete electric light and power system. As the newspapers reported, this was made clear to everyone who came to Menlo Park. The lamp was only the centerpiece of an array that included dynamos, switches, fuses, distribution lines, regulators, fixtures, and even a sewing machine motor. For the electric light to work at all, some kind of system like that put together at Menlo Park was necessary. For the light to work *commercially*, however, an entirely different level of systems problems had to be confronted. Of course, the effort at Menlo Park had always been guided by economic considerations; sometimes, as in the decision to develop a high resistance lamp, they were critical in determining the path of research. But by the end of 1879, with a practical lamp a reality, questions of the economy of the central-station electric light and power system came to the forefront.

There were important differences between the issues of practicality and economy. These differences, however, were largely of degree not kind, for the practicality of the electric light was just as much a relative matter as its economy. The standard for both was set by the gaslight, which, it should be remembered, was itself a technology in widespread use for only a couple of generations. People viewing the electric light asked whether its performance and convenience equaled that of gas and whether its cost was comparable. Edison embarked on the electric light project with a blind faith that, in the end, the answer to both of these questions would favor electricity. From time to time in the months before October 1879, Edison or one of his assistants took time to make rough calculations of the anticipated investment and operating expenses for an electric light system, often comparing the results with gas costs to come up with figures for possible profits (see Chapter 3). Now that the light

(opposite)
Upton's Optimism, 1880.
Francis Upton's optimistic portrait of the new lamp also reflected the feelings of his co-workers at the laboratory.

119

was a reality, the calculations took on a more serious character. For the first time, the cost equations could be filled in with information about the actual power requirements of a workable lamp and efficiency of an appropriate generator. And, of course, now that the key technical features of the system were complete, the patience of investors could not be imposed upon much longer without concrete steps toward commercial development.

An early effort to apply the knowledge gained from the new lamp to the question of the proposed system's economy was an extensive set of calculations recorded in December 1879 or January 1880 by Samuel Mott, Edison's draftsman, and labeled, "Preliminary estimate of the cost of lighting 10,000 Edison electric lamps from a Central Station each lamp giving a light equal to 16 candles or taking the place of a gas jet consuming 5 cu. ft. per hour." Assuming that the average daily use of a lamp through the year was 5 hours, or 25 cu. ft. of gas, Mott figured that the annual gas consumption of a 16-candlepower source averaged 9,000 cu. ft. For convenience, the Menlo Park workers used the letter M for the electrical equivalent of the light produced by burning 1,000 cu. ft. of gas, so that each 16-candlepower electric lamp could be said to consume 9 M of electricity per year. A system for 10,000 lamps would thus have to supply 90,000 M annually (the equivalent of a gas supply of 90 million cu. ft.). Tests with the carbon horseshoe lamp showed that 8 lamps could be lit for each horsepower (hp) applied to the generator, so that a station serving 10,000 lamps would require 1,250 hp. Initial investment for such a station, delivering 90,000 M per year, was estimated at $170,900, or $1.90 per M in Mott's terms. Annual depreciation was figured at $6,425, or 7.14 cents per M, labor costs at 13.2 cents per M, fuel at 11.7 cents per M, water at 3.9 cents per M, and rent, taxes, and insurance at 11.1 cents per M. The total of 47 cents per M was then figured as the actual cost of supplying consumers with electricity, an annual cost of $42,300 for a 10,000-lamp (90,000 M per year) central station.

If power could also be sold by the station during the day, additional investment and operating costs would be very small, but profitability would be much enhanced. Mott estimated that a 10,000-lamp station should be able to market 400 hp for about 10 hours a day for "small shops, elevators, leather sewing machines, pumping &c." The return on investment expected from such an establishment was quite respectable, ranging from 11% for light and power sold at 50 cents per M to 90% if charges were $1.50 per M (with some power sold at a lower rate). Mott concluded by comparing his figures with averages for gas: an investment of $9.00 for each 1,000 cu. ft. of gas capacity vs. $1.90 per M of electricity and operating costs of 85 cents per 1000 cu. ft. of gas supplied to the consumer vs. 47 cents per M of electricity. The economic calculations painted a rosy picture indeed for Edison's electrical system.[1]

With typical brashness Edison assured his investors that such calcula-

Cost Calculations, c. December 1879–January 1880.
Knowledge gained from performance of the new lamp stimulated calculations to determine the economy of the Edison system.

tions left no room for doubt about how his system would fare in competition with gas. When, at the end of December, for example, the secretary of the Edison Electric Light Company, Calvin Goddard, reported that a stockholder's friend had taken figures from newspaper accounts of the light and found gas cheaper than electricity,[2] Edison took time from hectic preparations for the New Year's Eve demonstration to reply:

Your favor of the 27th was duly received. I am preparing full statistics, etc. as to cost and I cannot now give time to refuting the statements of your gas friend, but briefly I would mention that no company in New York or in America can make and put 1,000 feet of gas in the holder for 35 cents, that there is 15 pc leakage chargeable to the 1,000 feet before it is delivered. That it does not cost 1 cent per hour per h.p. if I do not charge it to depreciation and interest, but only ½ cent per h.p. That I give a jet equal to 5 foot of gas burnt in Sugs Standard Argand. That if I run a 250 h.p. engine all night it will supply lights all night or if there is not enough lights to absorb 250 h.p., but say 100, then the Corliss cut-off works and there is just so much less coal burned. If you go into the delivery of gas then we have ½ the plant or interest 4 pc depreciation to 12 pc gas—no leakage to 15 pc gas and a chance to use our immense plant 10 hours of the day for selling power which is then sufficient in New York to pay interest on the plant and make the light for night for nothing. There are many other things which I could mention but I cannot at the present spare time.[3]

As so often in the past, Edison's pronouncements to both his backers and the public were wildly overoptimistic but served to keep investors' faith up and the public interest strong.

The obvious usefulness of this public optimism did not mean that it was gained by guile—in those months following October's breakthrough, the faith of the men at Menlo Park was very strong that the light they had discovered was all they had hoped for, and thus was competitive with any other. Even Francis Upton, who tried to be less cavalier about the difficulties of their undertaking, was full of confidence that the Edison system would be a profitable reality in a matter of months. He wrote to a friend in late January 1880:

I have been figuring during the past week on some estimates and they all show that we are going to make enormous profits at the present prices of gas, and fair profits at the present cost of gas. The lamps are durable as far as we can judge and no new troubles show themselves. I should say it is a good time to buy stock. There is no doubt of our going into New York in the course of six months. The patents are going to be excessively strong.[4]

Upton's December calculations were almost identical to Mott's, yielding a daily cost per M of 43.7 cents as compared to Mott's 47 cents.[5] In January, however, it became apparent that much more needed to be known about the dimensions of the task they were setting for themselves before Edison and his co-workers could move ahead. In particular, the extent and cost of the conductors supplying electricity from the central station to lamps and motors was beginning to loom as a crucial concern.

Charles L. Clarke was one of the many young men spurred by the excitement of Menlo Park in late 1879 and early 1880 to discover for themselves what was happening in the inventor's wonderful workshop. Clarke's major distinction from the others was that he was a good friend of Francis Upton, a classmate from Bowdoin College, and, like Upton, had studied in Germany. Thus, when Clarke joined the ranks of the Menlo Park workers, he was given the task, under Upton's guidance, of calculating the dimensions of conductors needed for a central station system. Beginning in early 1880, he filled several notebooks with figures and tables relating the sizes of conductors, the area to be served, and the cost.[6] These calculations were guided by principles Upton laid out for him. In one of his own laboratory notebooks, Upton spelled out "general laws" for figuring the cost of conductors, such as:

The cost increases as the square of the distance from the central station.

1 Ohm may be taken as a fair estimate of the resistance of a conductor for 10 lamps.

$\frac{1}{10}$ of the energy will be lost in the conductor when all the lamps are on.

If a 200 Ohm lamp can be made in place of a 100 Ohm lamp the cost in a district will be ½ for conductors or the distance from the station may be 1.4 times as great so that the station will supply 2 times as many lamps with the same average cost for conductors. . . .

The cost per lamp in a given area is proportional to the size of the area.[7]

The underlying principles and assumptions of the calculations favored from the beginning the relatively localized system that the first New York station at Pearl Street was to exemplify. The close relationship perceived between the costs of the system and the size of the area served discouraged any notions of service areas extending beyond a mile from the generating station. The variables considered by Upton, Clarke, and everyone else at Menlo Park—lamp resistance, conductor length, conductor resistance—could not have yielded any other conclusion.

Because the success of their system seemed to depend so much on a high level of usage in a relatively small area, Edison and his colleagues always framed their development work in terms of concentrated urban use, specifically, in a portion of New York City. To be sure, there were other reasons for this. From the beginning, the Edison Electric Light Company was a New York City enterprise backed by New York money, led by New York men, and coupled to New York interests. In addition, the press that lavished such attention on Edison was largely a New York press, and the only way to hold their serious and sustained interest was to develop a New York system. Moreover, from the time of Edison's

Conductor Calculations, January 1880.

Charles L. Clarke, one of the new, university-trained staff members and a friend of Francis Upton from college days, calculated this table of conductor dimensions for the central station system. He was guided by principles Upton set for keeping line resistance low and losses tolerable as the distance from the generator increased.

Distance	Length of Conductor	½ Ohm Resistance				
		Diam.	Area. sq. in.	Weight per foot. lbs.	Total Weight. lbs.	Cost.
20 ft.	40 ft.	.029 in.			.103	
40	80	.041			.413	
60	120	.051			.928	
80	160	.058			1,650	
100	200	.065			2,577	
120	240	.072			3,711	
140	280	.077			5,051	
160	320	.083			6,598	
180	360	.088			8,350	
200	400	.092			10,309	
222	440	.097			12,473	
240	480	.101			14,845	
260	520	.105			17,421	
280	560	.109			20,205	
300	600	.113			23,195	
320	640	.117			26,392	
340	680	.120			29,793	
360	720	.125			33,400	
380	760	.127			37,215	
400	800	.130			41,236	
420	840	.134			45,452	
440	880	.137			49,892	
460	920	.140			54,534	
480	960	.143			59,380	
500	1000	.146			64,431	

Horatio Alger-like adventures in New York as a youth of twenty-two, Edison had identified himself with the metropolis and saw it as the key testing ground for exciting new technologies. The questions and calculations he and his co-workers occupied themselves with in early 1880 were therefore questions about and calculations for New York.

When Francis Upton listed in his notebook the inquiries that had to be made to plan the central station system, he wanted to know about New York:

How many feet of gas are consumed in various parts of N.Y. per consumer and an estimate of the average time of burning.

How much horsepower is taken and the price paid per horsepower up to 10 horsepower.

Cost of pipes laid in the street and depreciation.

Cost of management.

Cost of building lots or rent of same.

Cost of water for H.P.

Houses—estimated cost of introducing.

Cost of condensing as compared to high pressure engines.

Cost of laying pipes or wires in the street.[8]

He filled much of the remainder of his notebook (fifty-eight pages) with figures acquired through inquiries concerning the costs of gas and other forms of lighting in the city and the kind of expenses expected for an electric central station and distribution network.[9] The results never differed greatly from the early calculations by Mott; indeed, one set of figures was identical to Mott's, suggesting that Upton may in fact have been responsible for all of the estimates made in this period. The calculations for system size and cost made in early 1880 were the first serious efforts of what turned out to be a long series of attempts to work out the expenses and the profitability of the Edison system.

As the Menlo Park workers turned their attention to the complex task ahead of them, following the heady success of the New Year's Eve demonstration, the key components of the system were carefully scrutinized in the search for improvements. One system element that had been largely left alone for the preceding few months, but which now came in for intensive work, was the generator. Since the "long-legged Mary Ann" had emerged from the Menlo Park shops in mid-1879, Edison had felt that it met the challenge of providing an appropriate generator for his system in its essentials. As questions of long-term practicality and economics came to the fore, however, particular elements of the generator were studied further. As early as January 2, 1880, Batchelor began sketching designs for new commutators and commutator-brush arrangements.[10] The spur for this may have been the order given John Kruesi and his machine shop crew to construct at least six new machines of several different sizes.[11] The new construction provided the opportunity to experiment with possible refinements; the original design and the need for different size machines required the reworking of a number of elements, particularly brushes and armatures. Batchelor took on the commutator and brush work, Upton was responsible for armature experiments, Kruesi assisted both men, and Edison kept his hand active in all the generator work.[12]

A typical example of the cut-and-try approach that characterized all of Edison's generator work was a series of experiments carried out on brush arrangements in early February 1880. The Edison generator, as did all generators, sparked at the brushes. A notebook entry of February 8 described one search for a solution: "We found by turning the brush at right angles to commutator the spark disappeared, which led to the following experiments." The next few pages were taken up with drawings of various brush-commutator arrangements, involving different angles for the brushes, different widths of brushes, and different contact points between the brushes and the commutator. Each variation elicited a judgment, ranging from a bold "NG" (no good) to "very good and durable" and, finally, "OK—very good—Standard, Can at this angle (32°) be placed in any position all over the commutator." The extent of satisfaction with his solution can be measured by Edison's care to initial and date this final design for possible patent documentation.[13] It is not obvious from these experiments whether Edison appreciated the fact that a far more critical consideration in sparking was the placement of the brush contacts at the "neutral point" of the commutator. Since such placement was largely a matter of fine adjustment of brush positions, it may simply have been an obvious procedure for the Menlo Park experimenter, even without a theoretical basis.

Efforts to improve the armature design of the generator were not as straightforward, since the deficiencies of the original design were not so clear. Upton was given the responsibility to devise armature experiments and to formulate general considerations for armatures and generator design as a whole. In early February he recorded experiments with armatures treated in a number of ways, paying particular attention to the heat generated at the armature surface. With the assistance of Kruesi's machine shop, various armatures with cores of painted or laminated iron plates were constructed and tested. While the concept of eddy currents in the armature cores was never mentioned in the laboratory records, the Menlo Park experimenters must have had some understanding of the need to localize core currents and to keep the internal armature resistance as low as possible. (Eddy currents are currents induced in the armature core rather than in the outer wire windings. Currents in the windings deliver useful energy. Eddy currents waste energy as core heat.) The limited understanding of the theory behind their machines was underscored by the occasionally bizarre approaches taken by Edison and his colleagues in their experiments. Upton, for example, recorded Edison's design for an armature made of a long strip of iron, 0.032 inches thick, wound in a spiral with paper insulation between layers.[14] Kruesi constructed such an armature of 80 feet of iron and judged the result "very bad."[15] Upton's results with an armature made of iron plates 0.037

inches thick and oxidized on the surface for insulation were much better—a load of 75 lights for 3 hours at 95 volts raised the core temperature to only 100°C. He noted that "this is the best ever done in the world and brings us home with this size of machine."[16]

Upton's contribution to generator development in this period went beyond executing Edison's experiments. As he was inclined to do, he tried to spell out the "laws" governing generator construction. His basic formulation was:

> The best machine is that which with the least amount of energy consumed on the magnets will give the greatest E.M.F. on the smallest resistance in the armature so long as the spark on the commutator may be kept under. Interest on investment, friction and magnetic friction being made as small as possible.[17]

Upton elaborated on these considerations in subsequent notes:

Brush-Setting Experiments, February 1880.
Edison followed his usual cut-and-try approach to generator development in experiments to determine the proper setting of commutator brushes to minimize sparking. It was critical to place the brush contacts at the "neutral point" of the commutator. Edison's appreciation of this point is not clear. Since proper placement was largely a matter of fine adjustment of brush positions, it may have been simply an obvious procedure for the Menlo Park experimenters, even without theoretical basis.

The minimum cost of the magnets is when the interest on the amount invested is equal to the cost of the energy consumed in them.

The friction of the bearings must be made as small as possible.

Magnetic retardation may be largely avoided using very soft iron.

Local action [eddy currents?] is in the Edison machine of no consequence.

Spark on the commutator may be made very small by placing the brushes endways and widening the commutator.[18]

Additional calculations showed examples of how reducing the armature resistance or increasing the size of the magnets could be justified as long as the additional investment was balanced by increased output or lower heat loss. It was a firmly established part of Upton's outlook—and Edison's—to specify clearly the relationship between technical and economic considerations. The eventual success of the Edison system was in no small measure a product of the clarity with which the men at Menlo Park understood this relationship, in individual instances as well as in the general scheme of things.

Working out the myriad details of the full-scale Edison system was to be the major preoccupation at Menlo Park during 1880, but the greatest single object of attention at the beginning of the year was still the lamp itself. The displays of the last days of 1879 had been truly great public successes. They had also, however, been intense field tests for the carbon paper horseshoe lamp and showed Edison that the lamp was still not all he wanted it to be. Correspondence and laboratory notes of that winter do not include a systematic critique of the lamp's performance, but problems were apparent in the tests and experiments that occupied the laboratory without a break from the moment work resumed after New Year's Eve.

Hardly any feature of the lamp escaped Edison's critical scrutiny in this period. His chief concern was the lamp's longevity, and life tests begun in late October continued. By late January one lamp had burned for more than 550 hours and several others were approaching 500 hours.[19] As gratifying as this was, the occasional long-lived lamp did not represent the dependability Edison felt the system required. Too many other lamps broke after much shorter periods, and the influences determining lamp life were not well understood. Some problems were in fact obvious; the cracking of the glass bulb at the point where it was closed around the lead-in wires was a continual headache, for example. Other difficulties were harder to pin down, such as the clouding of the inside surface of

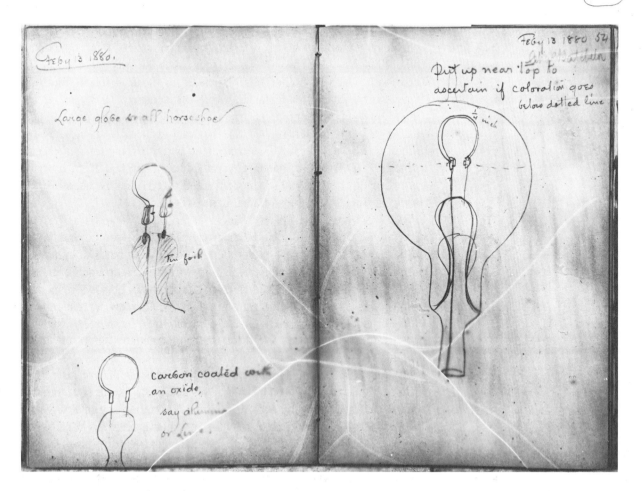

Feby 3 1880.

Large globe on aff horseshoe

Tin foil

Carbon coaled work
an oxide,
say al...
or f...

Feby 13 1880 52

Put up near top to
ascertain if coloration goes
below dotted line

¼ inch

some bulbs after a short period of use, and the unreliability of the fila-ments themselves, some lasting for days or weeks, others, identical in preparation and appearance, breaking at the slightest shock or giving out quickly when lit. The attack on these problems differed little in character from the months of work that preceded the carbon filament lamp, consis-ting of a broad range of experiments guided by empirical considerations and the occasional, sometimes crucial, insight.

There were, however, some differences in the working environment. The new blood attracted by the success of the carbon lamp was one source of change, for it allowed Edison to assign particular experiments to an individual who could devote himself entirely to the task at hand without diversion to other, momentarily more urgent, problems. Young Albert Herrick, for example, was given the task of keeping records of the performance of test lamps installed around Menlo Park, and Wilson S. Howell had the job of improving the operation of the vacuum pumps, now kept going at all hours evacuating experimental bulbs.[20] The widely pub-licized triumph at Menlo Park attracted not only ambitious young men

**Lamp Experiments,
February 13, 1880.**
Work at the beginning of 1880 focused on improving the carbon lamp. Some problems were difficult to isolate, such as the subject of this experi-ment—clouding of the inside surface of some bulbs after a short period of use.

seeking exciting new paths to success but also the attention of experts in one field or another who wrote to let Edison know what they could do for him. Edison's concern over some lamp problems was reflected in the care with which he sometimes responded to these unsolicited letters. One Albert Taylor, for instance, wrote on January 3, 1880, to recommend for lamp filaments a variety of bristol board made with the finest kaolin (a clay). Edison replied that he wanted board with no kaolin or other nonfibrous fillers.[21] This contrasted with his reply to a suggestion sent in by G. Haines on January 6 that Edison might find useful a bristol board made from linen. Edison instructed that Haines be asked to see about getting 100 pounds of linen bristol board.[22]

An interview with Edison that appeared in newspapers across the country on January 16 quoted him as saying that one source of difficulty in perfecting the light was the tendency for his glass bulbs to break, especially at the seal around the lead-in wires. This brought a rash of letters offering solutions and personal assistance. Edison's responses showed his worry about the glass problem. While a few correspondents were brushed off with polite explanations of why their suggestions were not useful, others were asked for more information. When one Philadelphia writer, describing himself as "a practical glass blower," said that he thought he could help out if he were sent a sample bulb or given train fare to Menlo Park, Edison instructed his secretary, "Grif, send a return ticket."[23] Laboratory records do not show whether any of this volunteered help produced useful results. In other instances, Edison's responses indicated his satisfaction with particular features of the lamp. For example, Edison's answers to several letters received in January with advice on improving the vacuum in his bulbs were generally polite but rarely invited further correspondence. While Edison did not answer every letter he received, and the replies that did go out were usually prepared by Stockton Griffin, the Menlo Park secretary, based on a few notes by Edison, the apparent care with which he read the large amount of unsolicited mail, even at this busy time, was remarkable.[24]

One letter received in mid-January that Edison did not pass on to Griffin for a reply was from the mathematician and astronomer Simon Newcomb, head of the Nautical Almanac Office at the U.S. Naval Observatory in Washington. Newcomb, one of America's most distinguished scientists, suggested that the light output of the carbon filament might be considerably improved if Edison found a more homogeneous and solid form of carbon than that provided by carbonized paper. Whether or not Edison had already reached this conclusion himself, the search for a better filament material in the months ahead was in the direction of carbon with a more homogeneous structure.

Newcomb's letter also gave a clue to how the American scientific community regarded Menlo Park and Edison's accomplishments there. Newcomb and Albert A. Michelson, then teaching physics at the U.S. Naval

A.—Geissler's Pump. B.—Sprengel's Pump. C.—McLeod's Gauge. D.—Geissler Tube. E.—Bulb containing Phosphorous Anhydride. F.—Bulb containing Gold Leaf. G.—Electric Lamp Bulbs. a.—Mercury Supply Tube. b.—Air Trap. c. and e.—Mercury Sealed Stop Cocks. d.—Discharge Tube. f.—Scale. f'.—Gauge Tube. g.—Connecting Tube. h.—Mercury Sealed Joints.

VACUUM APPARATUS FOR EXHAUSTING EDISON'S ELECTRIC LAMPS.

Light Bulb Evacuation, January 17, 1880.
This drawing in Scientific American *shows the pump Edison used to evacuate light bulbs. (Courtesy Smithsonian Institution)*

Academy and later to receive the Nobel Prize, were in the midst of plans for new measurements of the speed of light, one of the key physical problems of the nineteenth century, and Edison had invited them to set up their apparatus at Menlo Park. Newcomb replied that the lack of hotel facilities in the village posed an obstacle to their working there. "Otherwise," he wrote, "I should like very much to accept your kind offer, as I think the advantages of your laboratory would be very great."[25] Newcomb was possibly just being polite, and the image of Professor Michelson, who made the precise measurement of physical phenomena his life-long work, rubbing shoulders at Menlo Park with the ever-practical Edison is incongruous, but the fact that the prospect was even considered attests to the status achieved by the Edison laboratory in the highest circles of American science.[26]

The mail that came to Menlo Park was of little practical help to Edison. He knew that the problems of his light would have to be attacked as they had always been—through mobilizing the great resources at his command and concentrating them on what he saw as the most critical areas. In early 1880 the problem to which Newcomb had referred, that of devising a better filament, had a chief priority, and much of the laboratory's effort was brought to bear on it. In mid-January, for example, Edison put his chemist, John ("Basic") Lawson, to work carbonizing a wide range of materials in the furnace. It was projects like this, occupying only a few days, that gave Edison's research its reputation as a haphazard hit-or-miss activity. Indeed, Lawson's list of tested substances included its share of peculiar samples, such as flour paste, leather, macaroni, sassafras, pith, cinnamon bark, eucalyptus, turnip, ginger root, and a variety of gums, barks, and other plant parts or derivatives.[27] This list is reminiscent of Batchelor's in late October 1879, when he carbonized a variety of woods as well as such materials as coconut hair, cork, and celluloid (Chapter 4). Edison himself recorded experiments with a variety of woods during that winter, noting the effects of differences in methods of carbonization as well as in materials.[28] The tradition of wide-ranging, loosely directed experimentation was firmly established at Menlo Park, and the search for an improved filament was pursued in that tradition.

However, the importance of this kind of activity relative to other work at Menlo Park is open to question. Materials tested for suitability as a source of filaments consisted of perhaps several dozen substances, of which a few, like macaroni or leather, were bizarre, but most were raw plant fibers of one sort or another, clearly chosen according to logical plan. Furthermore, the time spent by Edison, Batchelor, and others on the random testing of materials was rather small compared to efforts made in other directions. For example, Batchelor spent much of January and February making filaments from paper and a few different plant fibers, particularly manila hemp, also varying filament sizes and modifying the carbonizing process.[29] Both Batchelor and Edison carried out nu-

Jan 24th 1880 43

Look Manilla hemp and knotted thread on ends so :— then pressed it between smooth vice and carbonized in carbon ◯ with hole in

Chas Batchelor

Bass Fibre —

Feb 24 1880 166

Chas Batchelor

make clamps so :—

Take Bass Fibre and drill holes so :— Bring the whole up in Vacuo and then cut off into clamps putting fibres in so :— It may be better to place the platinum wires in the holes before carbonizing

Manila and Bast Fiber Filaments, Winter 1880.

The search for a better filament material was guided by the perception that carbon with a more homogeneous structure was necessary. Many substances were carbonized, but the range soon narrowed to plant fibers such as manila and bast ("bass" in the notebook). Bamboo was eventually chosen for the filament used in the lamps sold commercially for the Edison system.

merous experiments aimed at improving the performance of paper horse-shoe filaments by paper treatments, such as boiling in sugar, and coating the filaments with metal oxide solutions much like those tried on platinum months before.[30] The image sometimes conjured up of a laboratory plunged into a wild search through Nature's storehouse does not reflect the true character of Menlo Park in early 1880. Instead, Edison and his co-workers sought solutions for the deficiencies of their lamp in a number of directions, sometimes guided by a clear perception of where the problems lay, and sometimes guided only by intuition, hope, or chance.

The search for a better filament material was matched by efforts to improve other weak points in the lamp. The original carbon lamp connected the filament to the lead-in wires with tiny screw clamps made of platinum. These were expensive, time-consuming to make, and not wholly satisfactory in performance. Hence, considerable time was spent devising alternatives that might be cheaper or more reliable. The clamps had to meet definite requirements: they had to join the platinum lead-in wires securely to the carbon filament, and they had to withstand the high temperatures of the glowing filament. In the course of his lamp work, Edison had encountered only two substances he could rely on to survive such high temperatures—platinum and carbon. Therefore, to minimize the use of platinum, the experimenters at Menlo Park naturally turned to carbon, seeking a form that could securely link the wires to the filament while tolerating the heat. Batchelor spent a good deal of time during February and March making carbon "clamps" (often just plugs or rings rather than true clamps) from various substances. Early results were inauspicious; for example, on February 1 Batchelor wrote:

> Made hemp fibres with clamps of plumbago, graphite such as used in lead pencils—they have got too much stuff mixed with them for us—seem to swell up and form gases or arcs which bust up the lamps. Clamps made for these were just a cylinder of graphite with hole to push on wire and holes in top for carbon loop.[31]

Batchelor subsequently described clamps made of "Wallace carbons," lignum vitae, ironwood, and coconut shell. The wood and shell clamps, designed to be carbonized along with the filaments, were the subject of particularly extensive experimentation.[32] For a while, Edison was convinced that carbon clamps would prove an effective and inexpensive substitute for platinum—hence the extended description of their construction in his caveat filed with the U.S. Patent Office on March 20. But by the end of March, he had to acknowledge that he had not found a truly satisfactory replacement for platinum or platinum-iridium clamps.[33]

Besides tackling the practical problems of the carbon filament lamp, the Menlo Park workers felt a need to understand better the general principles governing the behavior of the lamp. Efforts to acquire this information were closely linked with both lamp research and plans and cal-

Designs of Filament Clamps, February 3–4, 1880.
The search for a better filament was matched by efforts to improve other weak points in the lamp. In the original lamp, filaments and lead-in wires were connected by tiny screw clamps made of platinum, expensive and time-consuming to make. Edison turned to carbon, the only other substance he had successfully used in his lamps, to find a cheaper alternative.

culations for a full-scale lighting system. Once again, it was the scientific Upton who took up the task of laying out the basic physical relationships that determined lamp behavior and efficiency. He sought to define the links between the surface area of a filament, its resistance, the energy it consumed, and the light it produced. Upton's concern with the surface area of the filament was a holdover from Edison's very early conclusion that he should minimize the surface area of the incandescing element, primarily to reduce heat loss. At it turned out, later lamp makers did not

**Carbon-Horseshoe Lamp
without Base, c. 1880.**
*The lamp photograph clearly shows
the platinum screw clamps and lead-
in wires.*

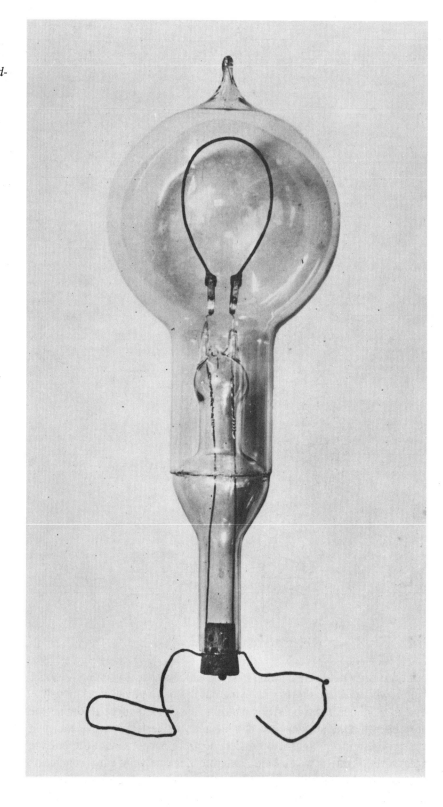

find it useful or necessary to make the surface area of filaments a major consideration. Thus, Upton's observation that "the light from the same energy is inversely as the surface from which it is given off" was not to be of practical importance. However, other relationships presented by Upton were pertinent, as exemplified by his statement:

A 200 Ohm lamp to give off the same light as a 100 Ohm lamp with the same surface will require 1.4 times the E.M.F. of a 100 Ohm lamp. That is, if a 100 Ohm lamp requires 87 volts a 200 Ohm lamp will require 121.8 volts. A 400 Ohm lamp will require 2 times the E.M.F. or 174 volts.[34]

This amounts to Upton's statement of the basic relationship between the power (W, or wattage) consumed by a lamp, the voltage (V) of the circuit, and the resistance (R), which can be expressed as $W = V^2/R$. This is a form of Joule's Law, determined by James Prescott Joule in 1840, for the heat produced by an electric current. It is curious that Upton, having had graduate training in physics, should express Joule's Law in such primitive terms. It may be that the implications of these basic physical principles for the electric light were still being worked out even at such a late stage.

A key consideration in Upton's work, as shown in both his studies of the lamp and his calculations for the generating station, was the efficiency of the carbon lamp. The question of how much light Edison was getting for the energy put into his generators was a question not only of great practical concern at Menlo Park but of general scientific and technical interest. In March, Professors C. F. Brackett and Charles Young of Princeton visited Menlo Park to run tests on the generators and lamps, carefully measuring the steam horsepower fed to the generator, the electrical power produced, the number of lamps lit by one machine, and the light output of the lamps. Their results, compiled in a detailed report submitted to Edison at the end of March, must have been gratifying to the Menlo Park researchers, for they confirmed the results Edison had been reporting.

The generator was judged to convert 86.7 percent of the input mechanical work into electrical energy. When armature losses were subtracted, 82 percent of the supplied power was available for use. A single lamp was found to consume an average of 0.0777 horsepower (hp), equivalent to 58 watts, so a single horsepower could operate 12.87 lamps. Thus, with an output of 4.717 hp for the Edison generator, 60 lamps could be lit by one machine. To make their tests comprehensive, Brackett and Young also measured the emitted light with a photometer, finding a peak illumination of 14.6 candlepower, well within the range of the gaslight standard.[35] Edison could not have hoped for a fuller validation of his own claims for the light.

Hard on the heels of Brackett and Young came two even more distinguished scientists to run very similar tests—Professors Henry Rowland of Johns Hopkins and George Barker of the University of Pennsylvania. Barker was a close friend of Edison and had followed with interest the work on the electric light from its beginning. Rowland, at thirty-one already one of America's most accomplished physicists, had always had an avid interest in electricity and at this point in his career was especially concerned with calorimetry (the measurement of quantities of heat). It is therefore not surprising that his and Barker's work "on the efficiency of Edison's electric light" (the title of their subsequent article in the April 1880 *American Journal of Science*) used a calorimeter to measure directly the lamp's power consumption. The most significant outcome of Rowland and Barker's experiments was not confirmation of Edison's claims for his lamp's efficiency but rather the observation that efficiency, expressed in terms of candlepower of illumination produced per horsepower, was directly related to the temperature of the filament—the hotter the filament, the more light it gave off, in terms of both absolute emission and relative energy consumption. Thus, the efficiency of two bulbs lit to a mean brightness of 11.6 candlepower was 109.0 candlepower per hp; the same bulbs at a mean brightness of 31.2 candlepower yielded 204.3 candlepower per hp. The physicists concluded:

> The increased efficiency, with rise in temperature, is clearly shown by the table, and there is no reason, provided the carbons can be made to stand, why the number of candles per horsepower might not be greatly increased, seeing that the amount which can be obtained from the arc is from 1000 to 1500 candles per horse power. Provided the lamp can be made either cheap enough or durable enough, there is no reasonable doubt of the practical success of the light, but this point will evidently require much further experiment before the light can be pronounced practicable.[36]

Edison was anxious to have the approbation of the scientists, for the apparent unsuitability of the light for commercial use after all the publicity in December gave rise to considerable skepticism in the press and elsewhere. Criticism in the newspapers struck a sensitive nerve, and the scientific confirmations were ammunition that Edison did not hesitate to fire back at writers and publishers. To the famous Chicago publisher, Joseph Medill, Edison got off a copy of Rowland and Barker's results with a letter that let loose the anxiety and annoyance that public skepticism was arousing in him:

> I send you tests made by Professors B. & R. Other tests are being made [by the] best scientific men. Your correspondent named Hall of N.Y. Tribune [Medill published the *Chicago Tribune*] came here

and stood about 20 minutes and then writes his long article, from inferences based on ignorance & misunderstandings. It is true that January 1st we put up about 80 lamps the first made, to ascertain the life of the lamps. They have all gone out. Their average life was 792 hours. The shortest 26 hours, the longest 1383 hours, a very satisfactory result for a 1st experiment. I have never concealed this fact when a lamp busted I let it remain where people could see it. My laboratory is open & free to sneaks, ignoramus detectives as well as gentlemen. I could have easily have made new lamps and replaced those destroyed had I wished to deceive the public.[37]

Edison pointed to the Rowland and Barker tests as proof of the economy of his system, but his exasperation as well as the stubbornness that was driving him finally burst out as he finished the draft of this letter:

After all these scientific tests are over, they will probably acknowledge the cheapness over gas, but they will say that the lamps won't last or that I cannot make them cheaply. [The rest of the draft is crossed out, but legible] In fact I am sorely puzzled to understand the motives of these attacks since I haven't cheated anybody but I am going to do all I said I would and more, too. Please don't publish this.[38]

The scientific community was far from being a unified ally of Edison, for shortly after the Rowland and Barker tests were sent off to Medill, there appeared in *Scientific American* the results of unauthorized tests by Professor Henry Morton and his colleagues at Stevens Institute in Hoboken. Using only one lamp (supplied by *Scientific American* because Edison had refused cooperation), Morton confirmed the basic characteristics of the lamp's operation. His test lamp yielded 16 candlepower when connected to 48 Grove cells (a voltage of 85–90V), had a resistance of 75 ohms hot and 123 ohms cold, and drew a current equivalent to 10.9A, which came to about 12 lamps per horsepower. Morton then went on, however, to calculate overall efficiency on the basis of the performance of well-known generators such as those of Brush and Siemens, which required $1\frac{2}{3}$ mechanical horsepower to supply a single horsepower of electricity. He compared the cost of this in coal consumption, about 5 lbs per hour, with the gas equivalent, about 25 cu. ft., and estimated that gas would produce some 110 candlepower. Claiming that the *effective* output of an Edison lamp was only about 10 candlepower, Morton thus asserted that the electric light cost about the same to operate as gaslight. To him this was, overall, a negative conclusion for the Edison lamp:

If each apparatus and system could be worked with equal facility and economy, this would of course show something in favor of elec-

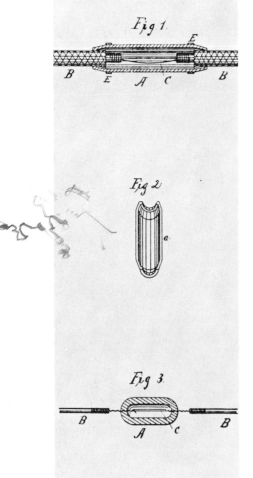

The Edison Fuse, May 4, 1880.
The "safety wire," or fuse, was an important feature of the Edison system and proved advantageous when the system was considered by the insurance industry.

tric light; but when in fact everything in this regard is against the electric light, which demands vastly more machinery, and that of a more delicate kind, requires more skillful management, shows more liability to disarrangement and waste, and presents an utter lack of the storage capacity which secures such a vast efficiency, convenience, and economy in gas, then we see that this relatively trifling economy disappears or ceases to have any controlling importance in the practical relations of the subject.[39]

While Edison could hardly have been pleased by Morton's statement and the negative attitude it reflected, he must have taken some satisfaction from the fact that Morton had reached his conclusions using generators with an efficiency of no more than 60 percent. The same calculations, using Edison generators with their efficiency of over 80 percent, would thus yield a far more positive picture of the Edison lamp's economy.

Even as the public debate over lighting economics was raging, Edison's men were hard at work putting the finishing touches on the first installation of the Edison system outside Menlo Park. During March and April 1880 the new steamship *Columbia* was docked in New York for outfitting, including installation of the first commercial Edison electric lighting plant. Henry Villard, head of the Oregon Railway and Navigation Company, the *Columbia's* owner and operator, saw the December demonstrations at Menlo Park and instantly became an enthusiast. In a matter of weeks, Villard (who, ironically, had at one time been a correspondent for Medill's *Chicago Tribune*) approached Edison with a proposal for installing the new light on the *Columbia,* then under construction at John Roach's shipyard in Chester, Pennsylvania. The *Columbia* afforded Edison the opportunity to put his light into commercial operation in short order under carefully managed conditions. The *Columbia* would also show off the electric light's real advantages over all alternative forms of shipboard illumination. Edison resisted all requests for lighting installations that he felt would distract from development of the full-scale urban system he envisaged, but Villard's enthusiasm and the belief that the *Columbia* offered a chance for a well-publicized effort with special popular appeal made it an opportunity Edison could not turn down.

The *Columbia* installation spurred the development of many auxiliary features of the system during the first months of 1880. The first lamp sockets, complete with key switch, were products of this period, since some means for securing lamps while in use was imperative for shipboard fixtures. The resulting design did not incorporate the screw socket but was a perfectly satisfactory approach that reflected awareness of the need to make the lamp easy to use by everyone.[40] Another item that had to be completed for use on the *Columbia* were "Safety wires," or fuses, that were to be the main protective elements of the system. In mid-April, Upton recorded a short series of tests on various gauges of

lead/tin w[...] how reliably
they melte[...]

The shi[...]amos—the
standard M[...]each—with
one dynam[...]e also the
Menlo Parl[...]filaments.
Lamp prod[...]each lamp
had to be t[...]ht outputs
were record[...], [...] done as well as Edison could manage, was necessarily chancy, and he resorted to a number of crude expe-
dients, such as rough cotton insulation for the ship's wiring. Francis Upton spent much of his time that April overseeing the electrical installation
and, when it was done, was anxious but hopeful, as shown in a letter
home written May 9:

**Steamship *Columbia*,
May 1880.**
*The first installation of the Edison
system outside of Menlo Park was
aboard the steamship* Columbia,
shown here in a Scientific American
*engraving. It afforded Edison an op-
portunity to put his light into commercial operation under carefully
managed conditions, as well as a
public demonstration with special
popular appeal. (Courtesy Smithsonian Institution)*

**Proposed Electric Light
Installation on *Columbia*,
January 31, 1880.**
*A letter by J. C. Henderson, engineer
for the Oregon Railway and Navigation Company, shows the proposed
plan for installation of the electric
light plant on board the* Columbia.

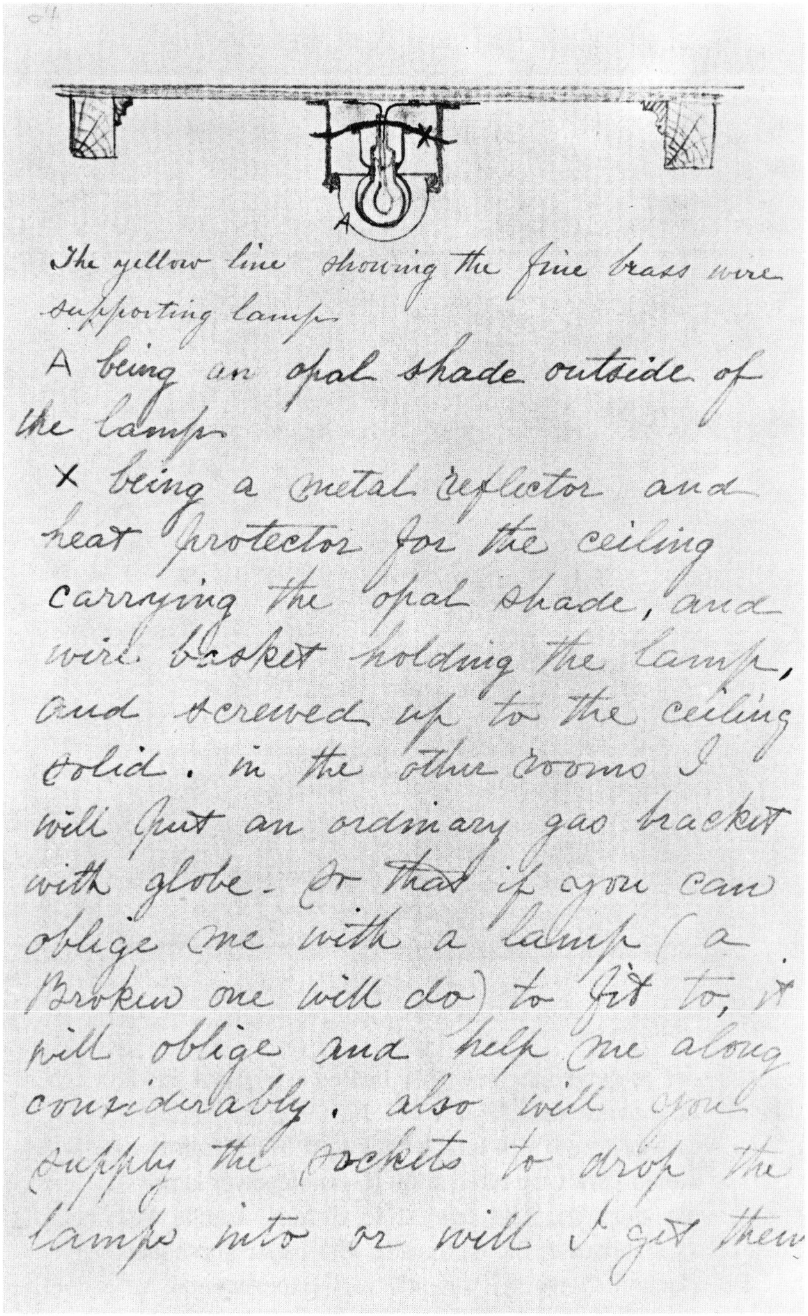

The yellow line showing the fine brass wire
supporting lamp.

A being an opal shade outside of
the lamp.

X being a metal reflector and
heat protector for the ceiling
carrying the opal shade, and
wire basket holding the lamp,
and screwed up to the ceiling
solid. in the other rooms I
will put an ordinary gas bracket
with globe. (or that if you can
oblige me with a lamp (a
broken one will do) to fit to, it
will oblige and help me along
considerably. also will you
supply the sockets to drop the
lamps into or will I get them

I am now through with the steamer Columbia, as it sailed yesterday
for San Francisco with the lights aboard. I hope that everything will
go well and see no reason to anticipate trouble. It will please the
Californians when it arrives and be of use to the vessel. There will
be in future a great many steamers to fit with the light and this will
be quite a profitable business in itself. [43]

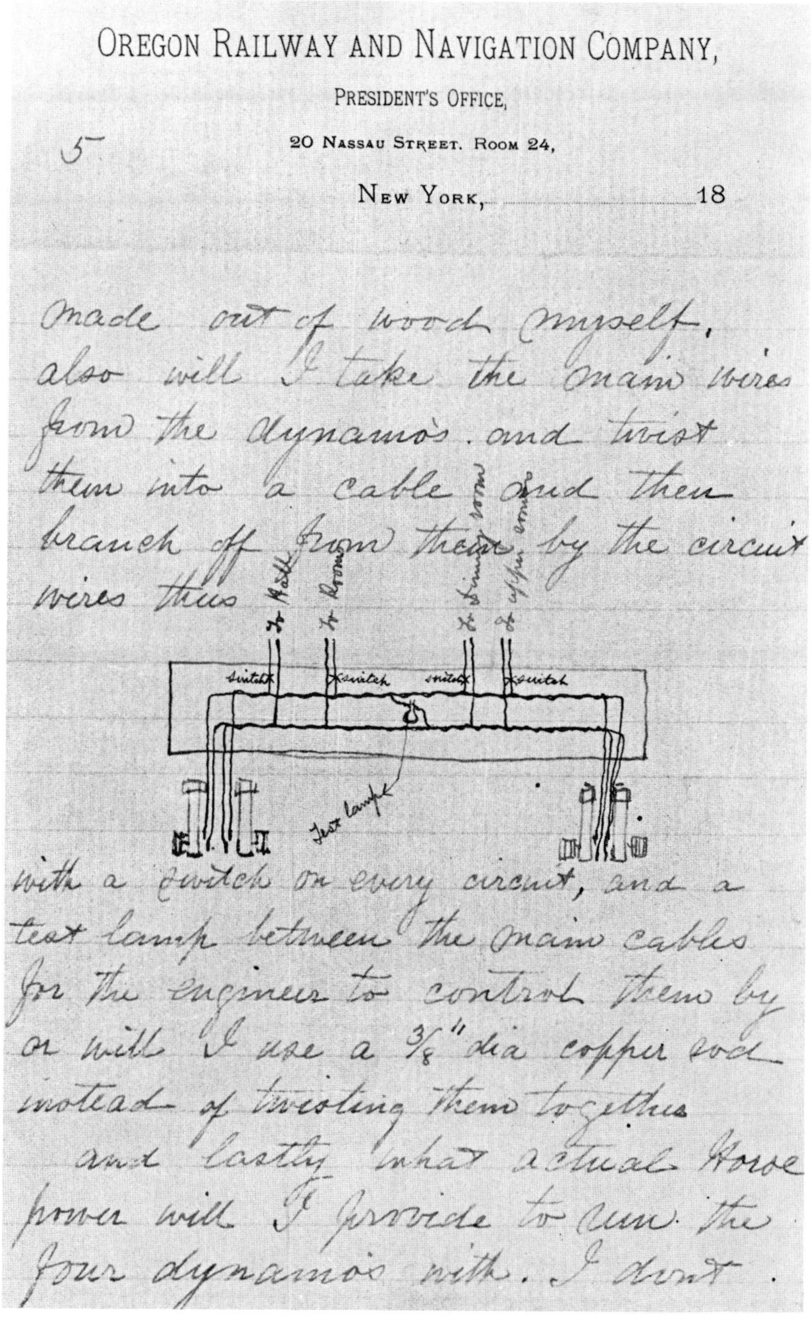

OREGON RAILWAY AND NAVIGATION COMPANY,

PRESIDENT'S OFFICE,

20 NASSAU STREET. ROOM 24,

NEW YORK, _____ 18

5.

made out of wood myself,
also will I take the main wires
from the dynamo's. and twist
them into a cable, and then
branch off from them by the circuit
wires thus

with a switch on every circuit, and a
test lamp between the main cables
for the engineer to control them by
or will I use a ⅜" dia copper rod
instead of twisting them together

and lastly what actual Horse
power will I provide to run the
four dynamo's with. I dont.

The outfitting of the *Columbia* was described in some detail in *Scientific American,* which observed of the Edison installation, "Certainly there is no place where a lamp of this character would be more desirable than on shipboard, where the apartments are necessarily limited in size and pure air is a matter of great consequence."[44] The *Columbia* project, while undertaken essentially as a favor for Villard, brought home an eco-

**Proposed Electric Light
Installation on *Columbia,*
January 31, 1880.**

nomic message that Edison was not quite ready to hear: The incandescent light would have an appeal and applicability in specialized, limited markets well before its place as the universal illuminant was established.

Indeed, as the development work was organized and pushed forward at Menlo Park in those first months of 1880, economic messages were coming to Edison with increasing insistence. By spring the technical problems that the lamp still presented were well-defined, and the dimensions of a future urban system were being calculated with care. More precise measurements of systems were being used to confirm cruder tests by the Menlo Park workers, providing assurances that the power needs and the operating costs of the carbon horseshoe lamp were well understood. But as time went on, the need to move economic considerations out of notebook calculations and into factories and actual installations became more obvious. In a sense, the development work at Menlo Park up to this point was still laboratory-oriented. To satisfy the public, investors, and himself, Edison would now have to bring development out into the open, into the creation of operating systems and a manufacturing facility to support them. It was to escape the distractions of installation and manufacturing that Edison had removed himself to the fields of Menlo Park in 1876, but the success of the electric light would drive him back into these activities and, in so doing, to the abandonment of Menlo Park itself.

The Menlo Park Mystique

Although Menlo Park has, with considerable justification, been called the world's first industrial research laboratory, it was not—in its essential organization—a prototype for those to follow. Later laboratories would provide an environment within which separate creative minds could work, stimulating each other and making common use of the facilities. For Edison, the laboratory was structured to serve only one creative mind, his own. As he said, "I am not in the habit of asking my assistants for ideas. I generally have all the ideas I want. The difficulty lies in judging which is the best idea to carry out."[1] This is not to say that others were completely stifled, but the rule in the lab was to carry out Edison's specific instructions first, pursuing other work if there was time. And without the master to provide direction, as happened from time to time, the pace of action dropped precipitously. Thus, Upton reported in a letter home that "One thing is quite noticeable here . . . the work is only a few days behind Mr. Edison, for when he is sick the shop was shut evenings as the work was wanting to keep the men busy."[2]

To keep an enterprise going under such circumstances was no mean feat. The small, intimate laboratory of phonograph days expanded substantially in late 1878 and again in 1879 for the electric lamp enterprise. Three new buildings were constructed (the office-library, the engine house, and the photographic studio which soon became a glassblower's shed) and more people were hired. In 1883 Edison recalled that "the place was crowded with people; I had thirty or forty assistants, and sadly neglected my usual care in dating exhibits and recording all experiments . . . I really had not the time." He went on to compare the activity with an only slightly earlier time when he was working on the telephone and when "I only had two or three assistants. We had more time, and I did a great portion of the work myself, whereas with the electric light I had 20 or 30 assistants and things were going on with a great rush, and I could not make the records myself."[3]

The actual size of the work force at Menlo Park can now be estimated with some accuracy. Time sheets are available for the last five months of 1878, for three scattered dates in 1879, and for all of 1880. Additional information can be gleaned from various accounts and court records, and from the files of the Edison Pioneers. The data speak graphically of the level of activity in the laboratory. The number of employees, fifteen at the beginning of August 1878, began to increase in October, reaching about two dozen at the end of the year. This number held fairly constant through much of 1879 until the late summer or fall when new hires raised it to about thirty-five. The successful demonstration of the light at the year's end, and the new activities that followed, produced a rapid increase at the beginning of the new year. In February 1880, sixty-four people were working with Edison at Menlo Park, a number that held remarkably constant for the rest of the year. There was, of course, a lot of turnover. By the end of 1880 some 220 persons could claim to have worked with the Wizard on his electric lamp.

Edison, as is well known, worked long hours, preferring those after dark, and expected his men to stay with him. Stay they did, late into the night and frequently until daybreak.[4] As John Ott put it, "My children grew up without knowing their father. When I did get home at night, which was seldom, they were in bed."

Charles Flammer had a room

in the laboratory where he worked putting carbons in lamps. He "slept there nights, or whenever I got a chance, but it was very seldom I slept at night."[5] There was compensation, in the form of overtime pay (at the regular rate), but there was evidently a higher degree of motivation at work, creating an enthusiasm that George Bernard Shaw—briefly employed to promote Edison's London telephone operation—noted in his preface to *The Irrational Knot*. The American technicians who had come to London, wrote Shaw, "adored Mr. Edison as the greatest man of all time in every possible department of science, art and philosophy."[6]

Those who came to Menlo Park to engage in the light bulb adventure did so with a variety of backgrounds and for a variety of reasons. Upton, Griffin, and Jehl came because Lowrey sent them. Charles Mott came to join his brother Samuel, who had come because he wanted to learn the electrical business. Albert Herrick first visited the laboratory with his mother who was doing an article for *Century* magazine. He was only seventeen, a student, but at Upton's suggestion began work just before Christmas 1879 and stayed for what must have been a very educational year.[7] Boehm came in answer to an advertisement for a glass-blower.

Wages were generally in ac-cord with the market, though there was considerable flexibility. Some began by working for free. And some negotiated. Typical, perhaps, was Samuel Mott, who had briefly studied drafting and electricity at the Princeton School of Science and arrived at Menlo Park with a letter of introduction to Upton from one of Upton's former teachers. He later said he would have accepted anything, that he would even have paid Edison for the privilege of working. He was offered $5.00 a week, which he negotiated upward to a respectable $7.00.[8] Boehm, who was responding to a desperate need for glass-blowing talent in August 1879, talked the Wizard out of $20.00 a week, which was raised to $30.00 at the first of the year. Edison testified that he subsequently was able to pay much less to others of equal competence.[9]

The glue that held all of this together was clearly Edison himself. A loner who was uncomfortable in normal circumstances, he had developed during his years as an itinerant telegrapher a talent for easy rapport with workingmen. Motivation for the men came from several sources. There was respect; Edison worked harder, longer, and more effectively than the rest. At the same time he was a peer. When John Ott first saw Edison, he "was as dirty as any of the other workmen, and not much better dressed than a tramp. But I immediately felt there was a great deal to him."[10]

Also, the operation was structured so that he worked with the others. Because everything flowed from his inventiveness, he was naturally interested in everything that was being done and pursued all activities with a watchful eye and pertinent suggestions. He knew when to take a break. Often this was at midnight, with coffee, pie, a cigar, loud music on the organ, and a round of jokes. Edison once said, "I was very fond of stories and had a choice lot . . . with which I could usually throw a man into convulsions."[11] He was also an impossible prankster, able to liven up proceedings when necessary, and encouraged others to do likewise.

In a word, the laboratory was not just a workplace. It was a home away from home. For the bachelors in particular, living at Mrs. Jordan's boarding house, the laboratory was the center of their lives, to which they turned for a variety of home needs. It was not uncommon for a man, finished for the day, to linger in the evening to watch what the others were doing. When it came time for a midnight snack, the working crew would often be augmented by a number of extras, who would disappear again when work resumed.[12] Upton expressed the mood in a letter to his father about four months after he moved to Menlo Park: "I find my work very pleasant here and not much different from the times I was a student. The strangest thing to me is the $12.00 I get each Saturday, for my labor does not seem like work, but like study and I enjoy it."[13]

One can only speculate how long the spirit of Menlo Park could have survived, or how many people it could tolerate. The numbers during the lamp days were apparently still tolerable. But when Edison recreated the laboratory on a much larger scale in a more urban area, at West Orange, the essential character was no longer present.

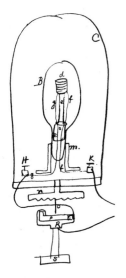

CHAPTER 6

A System

Complete

In the first months of 1880, the laboratory at Menlo Park completed the transition from a research establishment, devoted to discovering how to construct a practical incandescent lamp, to a development center, driven by the economic and technical requisites of a marketable system. In some ways, the activities were little different than they had been: a myriad projects pursued at once, a dozen or so workers assuming particular responsibilities, all under the general guidance of Edison himself. The problems wrestled with and the methods of solution also resembled those of years before; they were usually approached in an unremittingly practical way, broken down into discrete tasks that appeared to present a limited and manageable range of options. The laboratory's swollen staff, however, was now less the personal staff of a mercurial inventor than an organization serving the interests of an enterprise as much commercial as technological—the electric lighting system.

The changes at Menlo Park were not sudden or simple. It was still Edison's laboratory and, as such, reflected the liveliness and imagination of the man who once promised "a minor invention every ten days and a big thing every six months or so." Throughout 1880 the change in spirit and orientation was steady, driven by the technical demands of the lighting system. To understand what was happening, it is more important than ever to comprehend the entire complexity of the laboratory's operations rather than focus simply on the records of electric light experiments. Fortunately, just at this time, in March, a remarkable record of the day-to-day work of the people at Menlo Park was begun by Charles Mott, who had joined Edison's office staff at the beginning of the year. This account, part diary and part logbook, was apparently an effort to keep track of all the laboratory's projects and the activities of all its most important staff. Mott does not tell why he kept this record, but it was clearly not just for personal reasons. In the beginning, he cross-referenced many of his notes with indications of pertinent laboratory notebook entries. While this practice became less consistent as time went on, Mott's journal still provided a summary of the activities and ideas often recorded in more detail elsewhere. This careful record of daily activities may have been a task assigned to Mott to provide both a means of retrieving notebook data (a kind of narrative index) and additional ammunition for patent problems that might arise.

Whatever the reason for the journal, Mott's picture of Menlo Park's activities is far more comprehensive than can be derived from any other source. For about a year, from March 1880 to March 1881, he jotted down in small pocket notebooks his observations of what people were doing, who was absent (and for what reason), and what the laboratory notebook entries for any particular day covered. Supplementing his rough notes with conversations with laboratory workers, Mott entered his observations in a narrative journal. An extended look at this journal gives some flavor of Menlo Park's hectic pace as well as an occasional glimpse of the less serious side of life in what was to many the world's most exciting workshop.[1]

When Mott began his journal, the men at Menlo Park were involved with not only the electric light but other projects, some old, some new. The telephone work that had previously occupied Edison's attention, and which was closely tied to commercial efforts in Britain, was still a source of concern, and in March 1880 a few workers were set to improving the chalk-drum receiver Edison had devised to get around Alexander Graham Bell's patents. In a curious conjunction of efforts, one of the possible improvements under investigation was the application of an electric motor to the chalk drum, which had to be continuously rotated when in use as a receiver. Another project diverting some attention from the electric light was the continuing study of ore samples being sent to Menlo Park from all over North America. This study had begun in 1878 with the attempt to locate new and cheaper supplies of platinum at a time when Edison was convinced the expensive metal would be the basic incandescent element. Despite the fact that this was no longer the case, ore samples continued to come in and Edison continued to have them tested. The tests, however, were concerned less with the platinum than with the gold content of samples, and Edison's interest derived less from the needs of the light than the attraction a possible new source of gold has always had for ambitious men.

Closely related to this were the efforts being made to devise and perfect a magnetic ore separator—a device for using a powerful magnet to separate and concentrate various useful metals in low-grade ores. Mott's journal describes ideas for such a device in its entry for March 25—the first details of a project Edison was to return to often in the next years

until, in the 1890s, he poured himself and his fortune into a futile effort to revolutionize iron production in America. The magnetic ore separator, like almost all other side projects that surfaced in this period, owed its origins to the development of the electric light; the requirements of the lighting system ultimately governed the pace and direction of all activity at Menlo Park.

Mott's first journal entries cover some of the details of work already dealt with here in previous chapters. He recounted the visits of Professors Barker, Brackett, and Rowland and gave his own description of the calorimetric measurements of the light's energy consumption (March 14). His description of the construction of the laminated armature clearly spells out its purpose:

> (March 15) Cunningham finishing the *first* armature made from exceedingly thin sheets of iron (the thickness of tin) insulated with tissue paper, the last one put at work about a week ago was constructed on the same principle but with rather thicker discs say $\frac{1}{32}$ inches thick, those of the present make will contain about 600 discs

Menlo Park Laboratory Staff, 1879.
On the second floor of the Menlo Park Laboratory, with the staff around him, Edison is seated in the middle with a skullcap.

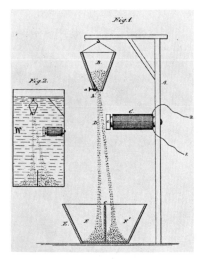

Edison's Magnetic Ore Separator, June 1, 1880.

Although the electric light system was the main project of the Menlo Park Laboratory during 1880, work also proceeded on others, some old, some new. Among them was Edison's magnetic ore separator, which grew out of his platinum search. His interest in mining later led to an extensive iron ore milling operation in New Jersey, incorporating such a separator.

of tin thickness and insulated as stated. It is found that this construction produces vastly better results than the thick uninsulated plates and does not heat to any injurious degree.

Mott also followed the work on experimental filaments in some detail, describing the devices made to cut and shape them from wood and the various approaches taken to improve methods of anchoring the filaments to clamps. The research particulars covered by Mott were, by and large, just the ones that might be expected, but now they can be viewed in a more precise chronological context.

Unlike his coverage of many technical matters, Mott's observations on the procedures and personalities of the laboratory are unique. Shortly after beginning his journal, for example, he noted a basic change in operations:

Sunday Mch. 21

Dean continuing work on new mould, but otherwise nothing doing in Shop or Laboratory. To night is the first Sunday night since I have been here and I believe for months before I came, that the Engine has not been running and work in the Laboratory proceeding the same as at other times, and it is now understood that night work for the present shall be discontinued. Mr. Batchelor's complaining of his eyes is the present inducement for making this change, which will undoubtedly prove advantageous to all who have from choice or necessity worked nights and slept and rested during the day—and a saving of considerable expense in night meals which have heretofore cost from $120 to $140 per month during the winter.

Even in his more typical dealing with technical work, Mott gives us a much more intimate sense of a day at Menlo Park than we can get from other sources. The day following the previous entry, for example, finds this record:

Monday Mch. 22

Dean finished mould for Plumbago ends for fibers and this evening Mr. Batchelor made two unsuccessful attempts to get the fibers out in tack [tact]. The mould however does its work very nicely and he will probably be able to get out some during tomorrow. Mr. Edison sketched another design for mold for same purpose and gave to Dean to make.

Have been running large armature all day for heat test—find it heats some but could not learn from Mr. Upton the number of degrees increase, not so much however I believe but what it will be able to stand.

Mch 18th 1880

New Mould for moulding plumbago ends on Fibres after Carbonizing

x x only held by steady pins

Piece on back of this to keep punches from pushing back all surface hardened and polished

ChasBatchelor

Made by C Dean

Mold for Attaching Filaments, March 18, 1880.
One method of securing filaments to lead-in wires used graphite ("plumbago"). This device was designed to mold graphite onto the ends of filaments.

Reported here that on Friday last Mr. Edison in New York City disposed of his relay to Western Union Tel. Co. for $100,000.

Profs. Upton and Jehl at work on Photometric and calorimeter tests probably for their own edification and practice as it is entirely useless to make any enquiries of them in regard to their investigations. Prof. Barker here a short time in afternoon.

Wilber here and said that during last week he filed for Mr. Edison six applications for Patents (one a day). I coppied [copied] applications for two weeks stay in Interference cases Edison vs. Dolbear & Edison vs. Dolbear vs. [blank]

It is through Mott's images of the busy patent lawyer, the frustrated technician, and the slightly snobbish academic that the Menlo Park laboratory comes to life.

Despite the continuation of older projects and the persistent intrusion of new ones, the laboratory in 1880 revolved around the electric light and its support system. By late spring the most important problems had defined themselves: (1) how to make the light bulb sturdier and more reliable, (2) how to make bulb production cheaper and suitable for factory operations, (3) how to determine the quantitative requirements of anticipated system installations, and (4) how to make the generator more reliable and more efficient, especially with large loads. Efforts to perfect the bulb continued to focus on finding an improved filament material. Mott's journal relates the progress of the Menlo Park experimenters as they attempted approach after approach in search for the ideal carbon form:

> (April 8) Those at work on the lamp and on carbons (Batchelor, Force, Mr. Edison, Flammer and some of the men in Shop) greatly interested in the efforts to devise suitable means and devices for reducing woods to sufficient small dimensions for carbonizing, and have been trying several different devices, Ott being of the opinion that a very fine keen saw will leave the wood smoother and in better shape for the carbons.

> (April 29) Wood loop cut from the thin worked holly milled by Force and cut after manner and in same former used for cardboard, carbonized by Van Cleve, were measured and put in lamps ready for pump, resistence 125 and 194 ohms.

> (May 14) Carbonization. Several moulds of Bast fibers were carefully prepared and formed around wood for carbonization, but the wood proved very detrimental, every one having been broken in the moulds during the process. Van Cleve is preparing some more for trial.

> (May 20) Carbonization. Van Cleve carbonized three moulds of bent wooden loops by securing the strips in slotted nickel plates; he got them out very nicely and in good shape. Bast fiber. Four of the Bast fiber lamps were measured and tested with current of 103 volts they gave from 30 to 32 candles and about six per horse power. They were connected to main wires in Laboratory and during the first hour three of them broke in the clamps and glass but the fiber in each instance remained in globe unbroken. Showing the fiber to make strong carbon but difficult to form good contact with.

These are only samplings of the day-by-day entries by Mott on the often frustrating hunt for a more durable and more reliable filament material.

Besides a continuous record of the search for a better filament, Mott's journal also provides a much clearer picture of the different activities pursued as part of the quest. For example, as each general kind of material was considered and tested for suitability as a filament source, special

Device for Cutting Wood-Loop Filaments, April 6, 1880.
As each material was tested for suitability as a filament source, special tools were often devised to properly shape it. Several weeks were spent in making a wood-milling machine to turn out wooden loops.

tools had to be devised to properly shape it. For weeks, Mott wrote of work done to devise a wood-milling machine for preparing wooden loops. On May 21 he noted the result:

Wood Milling. Dean is jubilant over his success today in working the cam milling machine with complete success and getting out about 100 loops of box & holly in excellent shape and in several cases sawing them so perfectly that the whole five loops were left joined at the thick ends, although the machine have been worked for some time with indifferent success. Today is the first Dean has felt satisfied with its working.

Other work related to the filament search involved developing better molds for the carbonization process, methods for extracting gum and resin from wood prior to carbonization, and devising other shaping tools. Mott confirmed that the number of different materials tested was not very large. By late spring, tests were limited to a few kinds of woods and bast fibers. While the experiments on woods continued through the spring, the usefulness of naturally fibrous material became more evident as time went on, and bast and other fibrous substances were the subject of the bulk of the lamp experiments by summer. One of the primary advantages of fibers was their superior stability after carbonization. According to Mott (June 12): "Mr. Edison observed that the Bast fibers shrink in carbonizing about 17 per cent, against 33% shrinkage in paper, woods, &c." It was natural, therefore, that further experiments should focus on other fibrous materials.

Bast fiber received most of the laboratory's attention during June, but persistent problems were encountered in the connections between carbonized fibers and clamps. On June 25 Mott reported that some loops were cut out of osier willow and palm leaf samples but turned out to be very fragile after carbonizing. A few days later some palm leaf loops were carbonized successfully, but the first lamps made from "palmetto" were not very promising. More were tried over the next week without markedly greater success. Finally, upon returning from a short Fourth of July vacation, Mott wrote that "A collection of Bamboo Reed and choice Bast have been obtained and some loops cut out but none yet put in the lamps to test (July 8)." On July 10, he reported the first tests of bamboo, but nothing spectacular in the results: "Bamboo cut from top or outside edge of fan. Resistance cold 188 ohms at 16 C. 114 ohms and gave 8.6 per horsepower." Performance of the bamboo in a lamp was somewhat like that of the bast fiber: (July 12) "A bamboo lamp tested Saturday (July 10) was put at 44 candles this morning, got very blue at clamp and lasted 1 hour 15 minutes." Bamboo was to be the climax of the search for a sturdier, more reliable filament material for commercial use, but its debut in the Menlo Park lab, while promising, was no more exciting than that of dozens of materials before it.

A few more weeks of experiments were required to determine that bamboo was indeed the material sought after. It is instructive to follow Mott's recounting of how this conclusion was reached. Soon after the initial tests, Batchelor, who was in charge of the carbonization experiments, figured out the shrinkage of the bamboo after carbonization (20 percent) and gave instructions for making a former for bamboo loops.[2] By this time the laboratory had devised a means for speeding up the crucial life tests for new filament materials. Instead of being run at standard operating voltage to yield around 16 candlepower (the desired light output for commercial use), lamps were put at higher voltages and tested at outputs of 40 candles or more. This procedure reduced testing time from hours or days to minutes. Mott describes one early bamboo test: (July 17) "Bamboo. Carbon of Bamboo with slight notch cut in one side set burning at about forty candles; in about five minutes the clamp on one side melted down forming a globule on the end of the wire and destroying the carbon. Lamp Number 1277 Book No. 57 Page 159." Further tests indicated the real superiority of bamboo: (July 21) "Average test on Lamps: From the lamps so far tested at 44 candles, the average life was taken and found to be, for Bast 6 Minutes, Calcutta Bamboo 17 Minutes and paper about three minutes. The Bamboo carbons were in many cases imperfect which has probably reduced the average for them, but which will be undoubtedly raised when proper precaution and care is used in selecting only perfectly cut and prepared carbons." At about this same time efforts were made to acquire a better quality of bamboo, with further gratifying results: (July 19) "Bamboo Pure. Some Pure Bamboo (genuine) was obtained and given to Bradley from which to cut some loops. The genuine is very fine grained and works nicer than the other." This "genuine" bamboo was later referred to as "Japanese Bamboo" and did indeed prove to be superior to the "Calcutta Bamboo" they had previously used. By August it was clear that they had what they wanted, Mott reporting: (August 2) "6$^{in.}$ Bamboo. Lamp burned 3 hrs. 24 mins. at 71 candles and gave nearly 7 per h.p.—the best lamp ever yet made here from vegetable carbon." From this day on, Mott made no more references to experiments on filament materials; "carbons," "fibers," and "loops" always referred to bamboo, and the tools devised for shaping them were made with that material in mind.

While efforts went forward to improve the performance and durability of the light, much of the work at Menlo Park was directed toward designing means to produce the light bulb on a large scale for commercial use. Although the problems of the electric light's operational economy—the cost of power consumed—and capital requirements—primarily for generators and conductors—were recognized early by Edison to be crucial elements in the design of the system, the basic cost of the lighting unit itself was not a major consideration until it became an immediate practical problem in the spring of 1880. Until six months before, after all, the lab was trying to perfect a lamp that depended not only on a complex mechanism for maintaining its filament but on the use of appreciable amounts of expensive metals like platinum and iridium. The unit costs of any of Edison's platinum lamps would have been horrendous, and his faith in their ultimate commercial possibilities rested on the beliefs that platinum could be had much more cheaply, that a properly designed lamp would last indefinitely, and that quantity production of any device would achieve significant economies. Edison was fortunate in never having to test these beliefs, for the carbon lamp that emerged in the fall of 1879 was marvelously simple and made largely of common and cheap substances.

This good fortune did not mean, however, that commercial production of the lamp lacked significant challenges. The search for a better filament material was, after all, spurred by the difficulty of handling cardboard-derived carbons in making lamps. But the filament was only part of the production problems that the lamp posed, and Edison wasted little time in trying to disprove skeptics who claimed that his fragile glass globes enclosing an unbelievably high vacuum could never be more than laboratory curiosities. The installation aboard the *Columbia* provided the first real impetus for the development of a lamp factory. Only a few days after he began making lamps for the *Columbia,* Edison tackled the most critical lamp production problem, evacuation. Mott recorded this start:

> (April 1) Mr. Edison tonight commenced experiment on pumps with the view of using single instead of double pumps as at present also of combining or arranging a large number in small space has on two lamps of fiber carbon, globes made tube shape.

The extremely high vacuums achieved in the carbon lamps by Sprengel pumps were their most extraordinary feature, and adapting evacuation into a factory process posed a unique challenge. Making lamps in the laboratory required hours of close attention to the modified Sprengel pumps, which were prone to breakage and faulty operation. To adapt the pumps to factory use, the workers at Menlo Park sought to simplify their construction and operation and to mechanize their action as much as possible.

Efforts to simplify the Sprengel pump, to make it more reliable and faster acting, were regarded as extremely important, as shown by the amount of attention they received that spring and by the litigation that followed a number of years later between Edison and his chief glassworker, Ludwig Boehm. Mott's description of this work gives a sense of how the Menlo Park laboratory attacked this kind of problem:

> (April 9) Boehm made glass cyphon [siphon] for experiment on the new single Springle [Sprengel] principle pump. . . .
>
> (April 10) Pump. Dr. Moses trying the new Springle drop pump with syphon attachment, as a whole finds it does not give results entirely satisfactory. Mr. Edison with much good reason attributes the partial failure to the cyphon attachment rather than to the pump itself.
>
> (April 12) Pump. It was noticed that in pouring the mercury in the globe reservoir, that the force carried more or less air into the

April 13th 1880

Trying Arrangement and joint for sealing on lamps on single Sprengel pump, devised by Dr. Moses.

Mc Leod gauge capillary tube on the top of tube a. Diameter of capillary tube is about 0.33 m. m.

Diameter 46 m. m.

Böhm

Experimental Vacuum Pump, April 13, 1880.

A drawing by Ludwig Boehm of a design by Otto Moses, an Edison chemist, shows one of several experimental pumps suggested by members of the laboratory staff and constructed by Boehm during 1880.

globe from whence it found its way into the cyphon and thence to the pump to obviate which Dr. Moses had tubes made closed on bottom with apertures on sides near the bottom so as to break the fall of the mercury and distribute it in the bottom of the reservoir with less force; found it to remove the difficulty of air in the cyphon but still the pump did not work as completely as desired.

Over the next week, Mott's journal describes work on the pump by Boehm, Moses, Upton, and Jehl. One aspect that comes through is the free, competitive spirit that sometimes reigned when several workers tackled the same task; another is the opportunity for invention some-

**Experimental Vacuum Pump,
April 13, 1880.**
*Francis Upton designed this pump
built by Ludwig Boehm.*

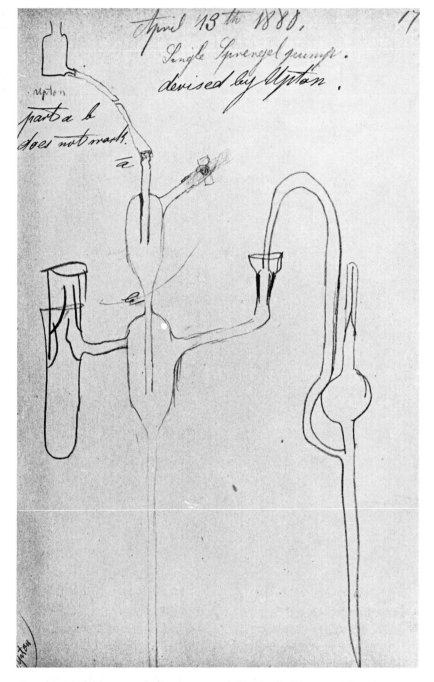

times seized by even the youngest of Menlo Park's crew (in this case, Francis Jehl):

(April 18) Pump. Francis had Holzier make pump after his design, which is very simple, free of stop cocks, takes but little room, and comparatively little mercury, tried in evening and pump vacuum up to one half of gauge tube obtained in little over an hour. He had also made a rubber stopper cup or socket in which to make the lamp connection without the necessity of using ground glass stoppers.

(April 19) Pumps. Francis put one wooden loop lamp on his pump in Dark room and got vacuum and heated up in two hours.

Dr. Moses with two paper loop lamps got good vacuum in five hours but had not heated. From these first efforts it would be difficult to say which of the pumps were best and quickest for getting vacuum, but the pump in use by Francis is much the simpler, cheaper, and occupies less room. Boehm making a new pump slightly different from either just mentioned.

(April 22) Pump. The single tube pump made by Boehm yesterday was started by Dr. Moses and he has succeeded in getting a good pump vacuum in 17 minutes and the pump appears to work quite satisfactorily. Francis also at work with his pump but so far as known has not timed it or obtained any data as to its merits.

(April 24) Pump by Francis again up and running with lamp on, and got vacuum in two hours. He had attached a device for flooding and removing the Mercury from the lamp cup or holder, made with small piece glass tubing attached to side of cup about one third down from top, to which is attached a short piece of rubber tubing running and attached to a small reservoir, bottle or other receptacle. By raising or lowering of which the mercury is forced in or drawn out of the cup, the rubber cork or stopper in the cup is cut sloping so as to deposit the mercury on the low side at the point of connection between cup and tube.

(April 25) Pumps. Holzier made six additional pumps of same pattern as one in use in dark room by Francis with new cup flooding device and put on frames with long slots in bottom and wooden slide fitted therein moveable up and down and to which is attached the tube receptacle of the gauge mercury.

(April 27) Pump. Francis claims good pump vacuum on his pumps in six minutes.

(May 3) Pump. John Ott taking one of the pumps from dark room apart and carefully taking dimensions of the tubing and all parts of the pump and making full sized diagram of same for use by glass blowers that they may make a number more of precisely the same size, caliber, &c. to ascertain whether they will give equally good results, and will be put in old factory and thoroughly tested as *the* pump.

It is perhaps not coincidence that on May 3, the same day Mott spoke of tests of "*the* pump," he also recorded that the first concrete steps were being taken to set up the lamp factory:

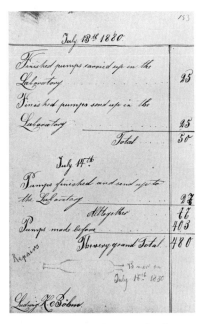

List of Vacuum Pumps for Lamp Factory, July 1880.

Ludwig Boehm built several hundred pumps for the new lamp factory at Menlo Park.

(May 3) Old factory. Along Rail road is being cleaned up and repaired preparatory to putting in pumps and lamp manufacturing apparatus.

Over the next few months Mott made periodic references to preparations of the old factory building (on the other side of the Pennsylvania Railroad tracks from the laboratory complex) for the installation of bulb making equipment. Manufacture of that equipment would, of course, be largely the work of the laboratory, so these same months were filled with that task as well. Here, as might be expected, pump construction was the greatest chore, even after Jehl's design had been adopted. Edison planned to install hundreds of the modified Sprengel pumps in the factory, and this necessitated the further design of machinery to keep them going, one example of which was described by Mott:

(May 4) Mr. Hornig making diagram of power pump to be set up and used in old factory for handling the mercury for the 476 pumps to be put up there, in which they are putting down floor.

(May 10) Mercury Pump. Mr. Hornig calculates 1000 lbs of mercury will be necessary to fill pipes and pump of capacity sufficient to 200 vacuum pumps. Mr. Edison decides to run the large Pump with electromotor instead of steam.

This "Mercury Pump" was one of the remarkable pieces of equipment designed at Menlo Park for large-scale light bulb manufacture. Operation of the laboratory vacuum pumps required constant attendance to keep the mercury reservoir at the top filled with the liquid metal. This was the job that had occupied Francis Jehl for so many months, and it was a slow, laborious, dull task. It was thus proposed that the factory be equipped with a large auxiliary pump that would mechanically fill the mercury reservoirs of the vacuum pumps. Edison's decision to drive this large pump with an electric motor drawing current from the generators that would provide light to the factory and surrounding buildings was the first step toward the industrial application of electric power.

Still other equipment and machines had to be designed for the lamp factory. These included an annealing oven to strengthen finished bulbs, carbonization ovens to bake quantities of filaments, and various shaping and milling tools for the preparation of filaments. One other device warrants a close look, for it reveals the appreciation by Edison and his colleagues of the important differences between the construction of experimental devices in a laboratory and the manufacture of consumer items on a large scale. The Menlo Park laboratory had for some time relied heavily on the services of skilled glassblowers, responsible not only for the often intricate glass work involved in making and modifying vacuum pumps and other laboratory apparatus, but also ultimately for the blow-

ing of the light bulbs themselves. Reducing the cost of bulb making would obviously require an alternative to the craftsmanship of the glassblowers. Mott described this development:

> (May 4) Lamp Mould. The metallic mould for lamp bulbs was secured on a permanent iron base and arranged with spring to open and void [evacuate?] and pully treadle for closing & works very completely. Pelzer here with two samples of lead glass tubing, and took with him one of the globes blown by Boehm as model from which to make wooden mould.

> (May 9) Globe Mould. Holzier blowed about one half dozen globes in the metallic mould and treated them with the acid solution. It may not require so much skill to blow in mould as off hand and therefore with cheaper labor, but as for appearance and time I think there is no advantage whatever in the mould.

Mold trials continued for a brief period, but in the end bulbs continued to be "free-blown." William Holzer (Mott's "Holzier") joined the staff in January 1880, leaving work in Philadelphia (see above, Chapter 5) where

Menlo Park Lamp Factory, 1880. *A large staff was employed by the Edison Electric Lamp Company at Menlo Park. Setting up the lamp factory, installing machinery, and testing equipment for making lamps occupied the workers throughout the summer.*

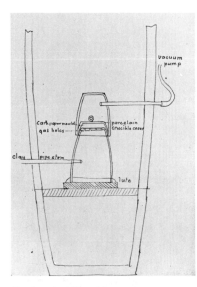

**Carbonization Furnace,
January 1880.**

*Otto Moses designed this furnace to
carbonize lamp filaments.*

he had lived for many years. Holzer may have been responsible for urging Edison to contact the Corning Glass Works, which sent a representative to Menlo Park, according to Mott, on June 21. Not long after, arrangements were made with Corning to supply blown bulb blanks (made from "pot glass") although it was November before the first shipment of 307 dozen blanks arrived at the Menlo Park factory (at a cost of $5.00 per gross).[3] The bulb blanks remained the principal subcontracted component of the Edison light bulbs.

Preparation of the lamp factory, construction of factory machinery, and the installation and testing of equipment occupied workers at Menlo Park through the summer. Construction of more than 500 vacuum pumps was a considerable job in itself, requiring the full-time attention of a group of glassblowers for several months. The large, electric motor-driven mercury pump turned out to be very difficult to design and build for reliable, continuous operation and, in those early days of electric motors, incurred the loss of considerable time and energy in fixing crossed wiring and burned-out armatures. Even seemingly trivial problems often took weeks to solve, requiring men who would otherwise be supervising factory installations to return to laboratory work. For example, it was discovered early that repeated use of mercury in the vacuum pumps tended to foul it with impurities that soon impaired the pumps. Thus, the Menlo Park chemists, Lawson and Moses, had to devise methods of cleaning the mercury in the course of its use in the factory. In another instance, soon after the adoption of bamboo as the basic filament material, it was observed that many bamboo loops emerged badly bent from the carbonization process. At first, Batchelor found a clever way of straightening them after insertion into the bulbs; Mott described it this way:

> (Aug. 4) Straightening carbons. Some of the longer carbons have at time bent over to one side vary much after they had been placed in the lamps and heated. To straighten them the lamp is placed between the poles of an electric magnet and the current turned on on the lamp. One pole attracts while the other repels the charged carbon and the lamp is placed between the poles in such relation that the polar action is utilized to straighten the carbon through the glass globe. I find that this "little dodge" has been worked for some time but today is the first I have caught them at it.

This expedient was, however, not suitable for the factory. Batchelor continued attacking the problem, and Mott's description of how the solution finally emerged gives another revealing glimpse of the Menlo Park style:

> (Aug. 12) Carbonization. Mr. Batchelor's experiments with different heats did not solve the problem or reveal the cause of the loops

bending over. A thorough discussion to night suggested the theory that it might be due to the fact that the way the present loops are cut and carbonized the pith and outside of the Bamboo always comes on the side of the loop, and to test the theory whether the pith side, being more loose and porous than the outside did not shrink more from the effects of the heat, than the more firm and compact outside. It was determined to change a carbon forming mould so that the widened ends might stand at right angles with the face and thus bring the pith either on inside or outside of the loop. The mould was then arranged and between twelve o'clock and morning Mr. Batchelor got out three or four, had them put in lamps exhausted and heated very high and to the gratification of all the loops remained erect and they justly feel that the problem is solved and that carbonization is now worked to a fine art.

Batchelor's own laboratory notes give another perspective on how this problem was attacked:

> The cause of the bending over of the loop after it is heated in vacuo I thought was due to insufficient heating in carbonization but after a series of experiments to determine that point we came to the conclusion that whether heated slightly or to a high temperature some of each bent whilst others kept straight.
>
> We then remembered that some bamboo fibres which were 4″ long and of which we made a great number almost all kept straight. We also remembered that almost all of these were put in the clamps edgeways instead of flatways. This led us to see that the way Bradley cut them from the cane, and the bending them flatways afterwards, would have the "*pith side*" on one face and the "hard-shell" side on the other face[;] unequal shrinkage of course must occur on two such faces and cause the bending.
>
> We now made a mould for carbonizing that would hold the fibre edgeways so. . . . From this mould we tried some on the pumps and they not only were perfectly flat themselves but did not change their upright position with the most intense heat we could get on them.[4]

The problem of the bent bamboo loops was just one of many small (but often crucial) details that had to be worked out as lamp manufacture was systematized. Mott's reference to the "thorough discussion" that followed Batchelor's initial difficulties provides a good instance of the cooperative team effort that was as characteristic of the laboratory as the independent and friendly competition illustrated by the work of Jehl, Boehm, and others on vacuum pump design. It is well to note that Edison's participation is not mentioned by either Mott or Batchelor (indeed, Mott records that Edison was absent all day), for, in dealing with such

Device for Keeping Bamboo Filament Ends Straight, March 18, 1880.
After the adoption of bamboo as the standard material, it was observed that many filaments emerged with bent ends from the carbonization process. This device was one method used in an attempt to overcome the problem.

details, the laboratory could function very well without Edison, particularly when the details were being handled by Batchelor, Upton, or another member of the team with substantial responsibilities. Batchelor's notes offer a revealing picture of the kind of reasoning applied in laboratory discussions—an empirical and yet clearly deductive approach that is the mainstay of any establishment devoted to solving technical problems. The "perspiration" that Edison said was ninety-nine percent of inventive

this! —
43

a thin plate of nickel lies under the fibre and has a side turned up — The weight lies on top of this plate with the fibre ends in between — The sides of this plate ~~too~~ also confine the body of the fibre to a smaller chamber thus making less liability to oxidization ————

Aug 4 1880
Chas Batchelor

Filament Carbonizing Mold, August 4, 1880.
Several molds were tested to reduce filament distortion or deterioration during carbonization. This design presses the ends flat and lessens exposure to air.

genius was effective because it was guided by alert and perceptive thinking.

Other obvious deficiencies of the lamp were attacked in a similar manner. The design of the clamps that connected the filaments to the lead-in wires presented very difficult problems that took many months to solve. The original platinum screw clamps were not only expensive and time-consuming to make but often performed poorly and Mott recorded with

**Clamp-Making Machines,
May 8, 1880.**
*Charles Batchelor attempted to speed
up the slow manufacture of clamps
with intricate machinery.*

Clamp Making Machine.

some frequency the appearance of a blue glow at the foot of many filaments in test lamps—a glow due to an imperfect connection between
filament and clamp. In the summer of 1880 nickel was found to be a suitable substitute for platinum, but clamp manufacture was still difficult and
performance no more reliable. After lamp manufacture began in the factory in the fall, the clamp problem became critical. Finally, in December,
the chemist Lawson was called in to see what he could do to eliminate

Part for Clamp Making Machine.

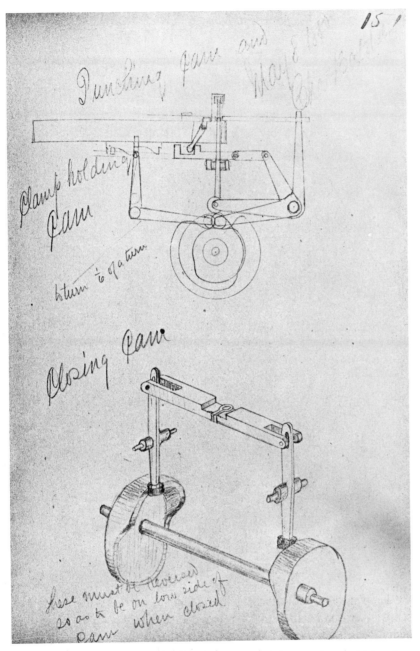

the need for clamps altogether. His solution was to plate the ends of the carbon filaments with copper and then join the plated filament ends directly to the lead-in wires by further plating.[5] Surviving lamps from the Menlo Park factory testify to its success.

Less urgent perhaps than the clamp, but still a challenge to Menlo Park ingenuity, was the problem of designing a base and socket for the commercial lamp. The laboratory method of connecting operating lamps—holding

Device for Attaching Filaments to Clamp, August 26, 1880.
This device was tried to solve the problem of attaching the filament to the lead-in wires by using screw clamps. Eventually, an Edison chemist, John Lawson, solved the problem by plating the ends of the carbon filaments with copper and joining them directly to the lead-in wires by further plating, thus eliminating the need for clamps altogether.

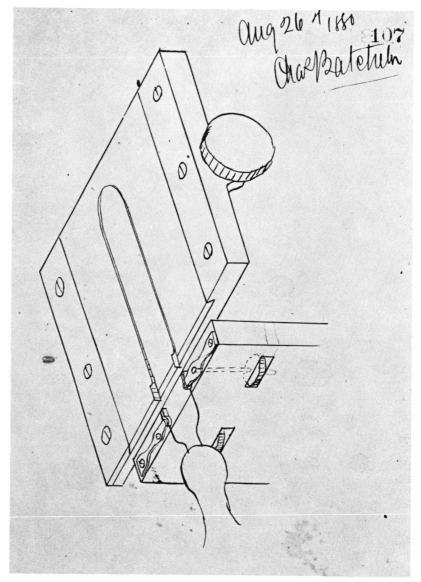

the lamp upright and attaching the lead-in wires to standard terminal screws—was not satisfactory for general use. The *Columbia* installation stimulated the first efforts to devise a method of connecting lamps that was both convenient and safe. In early 1880 the first lamp sockets were made, consisting simply of wooden receptacles with copper contact strips inside. Lamps had matching strips at the ends of the two lead-in wires. When the lamps were inserted into the sockets, the lamp strips would press against the socket strips and make the electrical connection.[6] This was good for upright installations, but was obviously not as generally stable and secure as might be desired. The final solution was, of course, the screw base and socket, which made its first appearance in

Lamp Socket   *Aug 25th 1880*   97

Chas Batchelor

The rings of brass xx to be fastened on to the inside socket

A

Early Lamp Socket, August 25, 1880.

This socket is similar to those used earlier on the Columbia *installation, which consisted simply of wooden receptacles with inner copper strips. When the lamps were inserted, matching strips on the lead-in wires made contact and completed the circuit. This was adequate for upright installations but not as stable and secure as desired. Charles Batchelor's notes suggest that the laboratory staff was leaning toward the screw socket design adopted soon after this drawing was made.*

the fall of 1880.[7] The evolution of this design in the commercial bulb can be followed through surviving specimens. The first screw base was of wood, with plaster of Paris used to attach the bulb to the base. Later, in 1881, the bulky wood was dropped in favor of more plaster of Paris, and still later that year the design was modified to place less stress on the base-bulb connection, and the modern socket form was complete (although materials continued to change over the years).

Since Edison was content that his long-legged Mary Ann dynamo design left nothing to be desired, and the *Columbia* installation showed that the laboratory could produce as many generators as needed for any particular system, the development of a new form of dynamo was not a task of the highest priority during early 1880. The difficulties of perfecting the lamp and establishing factory production were the focus of attention. Despite this, it is clear that Edison's conception of the commercial central-station power plant changed during the year, and a new goal for dynamo design was the most significant part of that change. The best statement

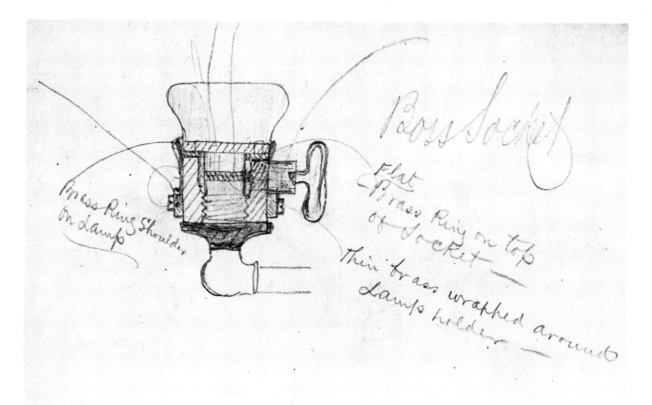

**Design for Screw Socket,
c. September 1880.**

*Edison's long-time associates,
Edward H. Johnson and John Ott,
were principally responsible for designing fixtures in the fall of 1880.
Their work resulted in the screw
socket and base very much like those
widely used today.*

of this goal appears in his *North American Review* article, "The Success
of the Electric Light," published in October. In it he describes the improved bamboo lamp and then outlines his envisioned power plant:

> Another important modification of the system, introduced since
> the latest authorized account of the light was published, is the sub-
> stitution of dynamo-machines for magneto-machines in the stations
> from which the electricity is to be supplied to the several districts
> of a city. Here, again, the change is entirely in the direction of sim-
> plicity and economy. Where before it was proposed to furnish a sta-
> tion with one hundred magneto-machines with a multiplicity of belts
> and shafting, we now make ten dynamos of 120-horse power, each
> worked directly by a 120-horse-power engine. We thus do away
> with a very considerable loss of power, and at the same time the
> outlay for machinery is very much lessened.

Like so many of Edison's public statements, this one cannot be taken
at face value. From the time two years before that he had experimented
with the machines made by Wallace and Gramme, Edison had always used
dynamos—generators having electromagnets—and not magnetos—gen-

erators with permanent magnets. Neither his long-legged Mary Ann nor any other machine he had seriously considered for use in his system had been magnetos, and yet he announced here his progressive step in adopting the dynamo principle. As for a 120-horsepower machine, however, Edison spoke more of a plan than a reality. Such a machine would be able to power more than a thousand lights, but no device built at Menlo Park up to that time could power more than fifty or sixty. In August 1880 Upton jotted down "calculations for 100 H.P. machine" but was clearly making plans for something far greater than had yet been built.[8] Most of the dynamo work during 1880 was directed toward building enough reliable machines of the standard bipolar design to meet the needs of the *Columbia* installation and then the larger needs of Menlo Park (the demonstration lighting system being erected around the village and light and power for the factory). These installations had precisely the pattern that Edison disavowed in his *North American Review* article, that is, an assemblage of enough sixty-lamp machines to meet total power needs.

By the end of the summer, however, his vision of a larger, more effective dynamo for a full-scale commercial system had taken hold. Because of the pressures of other projects, work on realizing this vision went ahead slowly. On September 7 Mott recorded: "Forgings of soft iron for the magnet arms or poles and top pieces for the large 100 horse power dynamo were [received]," and on the 11th: "Large Dynamo. Mr. Clarke finished the details and drawings for the large armature, and the heavy castings for bases of magnets were received." From that time until January 1881, work pressed forward on Edison's first large dynamo—the prototype of the "Jumbo" that was to emerge from his shops about a year later. The design and building of this machine posed considerable challenges for the Menlo Park crew, since it was their first attempt at a device that far surpassed laboratory-scale construction. Mott's journal entries for the fall and winter of 1880–81 allowed hardly a week to go by without some reference to the large dynamo project. Special implements and machine tools had to be made to fashion the large parts, particularly the armature and the commutator. Extensive discussions took place with steam engine makers, especially Charles T. Porter whose Porter-Allen engines represented the state of the art in the 1880s. One of the primary features of the machine that Edison envisaged was its direct connection to the driving steam engine, dispensing with the belt and pulley that had always been used with smaller dynamos. This required, however, that the characteristics of the steam engine be taken into account at every stage of the dynamo's design and construction. It turned out that the problem of building dynamos to work in such intimate unison with steam engines was never completely overcome by Edison, as evidenced by the difficulties he was to experience in 1882 with Jumbo dynamos installed in the Pearl Street station in New York City. As remarkable

Direct-Connected Dynamo, March 3, 1880.
This drawing is an early conceptualization of the larger, more effective dynamo Edison considered a key element of his full-scale commercial system. These machines, designed to be directly connected without belts and pulleys to the steam engine, were to be capable of producing 100 horsepower and supplying more than 1,000 lights.

Direct-Connected Dynamo with Porter-Allen Engine, 1880.
In the summer of 1880, work began on the construction of the large, direct-connected dynamo. The steam engine driver (at left) was designed by Charles T. Porter, whose Porter-Allen machines represented the state of the art.

as Edison's large dynamos were, most of the early Edison power stations, after Pearl Street, used smaller dynamos powered by belts and pulleys from independently operated steam engines. Direct shafting between dynamo and prime mover would for the most part have to wait until reciprocating steam engines were replaced by giant steam turbines a few decades later.

Despite the limited long-term significance of Edison's large dynamos, they reflect the engineering approach and logic of the Edisonian method as practiced at Menlo Park and therefore deserve careful scrutiny. The rationale for the large dynamo was suggested by Edison's *North American Review* statement that the new machine would "do away with a very considerable loss of power, and at the same time the outlay for machinery is very much lessened." As Edison saw it, the use of a large number of small, belt-driven dynamos multiplied the mechanical losses inherent in any extended linkage between steam engines and driven machinery. The simplicity of this logic, however, did not mean that he thought it was unnecessary to measure the actual inefficiencies involved.

While tests were occasionally carried out to determine the efficiency of the Menlo Park dynamos, it was on the eve of the completion of the prototype large machine that definitive tests were performed. The design of these tests and their results were recorded in detail by Charles Clarke, who by the end of 1880 was Menlo Park's chief expert on power engineering. Clarke compiled his observations of the test of January 28–29, 1881 into a report entitled, "Economy Test of the Edison Electric Light," which represented the most extensive technical assessment of the performance of the Edison system after its first full year of development.

The test was carried out on the completed Menlo Park light and power system first shown to the public the month before. This system incorporated the key elements of Edison's contemplated commercial central station installations, including underground conductors linking the lamps with the dynamos in a "multiple-arc" (parallel) distribution system. Clarke's test exemplified the technical precision typical of all that he did. On the evening of January 28, a new fire was started under the steam boiler at 9:07, the engine itself was started at 9:22, the full load of lamps was turned on at 9:26.5, and the engine was stopped at 9:21 the next morning after some twelve hours. Careful measurements were made of boiler dimensions and fuel consumption, steam engine performance, dynamo circuits, and lamp behavior. Lamps in the circuit included 399 full-size "A" lamps, rated at 16 candlepower with a resistance of 110 ohms, and 54 "B" lamps of half the size and resistance of the "A" lamps. Since the "B" lamps were in series-connected pairs, constituting a load equivalent to 27 "A" lamps, the total load was equal to 426 standard "A" lamps.

Three lamps failed at different times during the test, and Clarke calculated that the test load was reduced to 424.67 standard "A" lamps. His precision extended to calculations of the total power produced during the test, which he summarized in his report: "While the engine was running 27,674 lbs. of water were evaporated, the engine developing 20.054 horse-power for 4.5 minutes, and 83.67 horse-power for 11 hours and 54.5 minutes." Using these figures, Clarke obtained a value of 5.08 (424.67/83.67) lamps per gross horsepower. After correction for electrical and mechanical losses and for the effective number of lamps in the circuit, the result was 7.25 lamps per horsepower. This compared favorably with the measured value of 8 lamps per horsepower. The 0.75-horsepower discrepancy was charged by Clarke to "the increased friction of the engine and system of driving pulleys and belts" which, he noted later, would be largely eliminated by the proposed directly connected high-speed engine and large-dynamo combination. The report concluded that "the approaching completion of the large dynamo, it is hoped, will soon place it within Mr. Edison's power to give to the engineering public complete and reliable data showing even a greater increase in the economy of lighting by electricity."[9]

Three weeks after Clarke's tests, the large dynamo was nearing completion. On February 20, 1881, Mott jotted down in his pocket notebook (where he scribbled notes for later elaboration in his journals) that "all [is] in readiness" with the "Porter dynamo." The machine was run for a few minutes on the 24th, but problems appeared in the bearings. On the 26th Mott recorded that the "Porter-Allen run all day at about 300 [rpm] empty to smoothe her up." The next evening the giant machine was ready for a full-scale test, and Edison came down from New York to observe it. Mott's notes of the time indicate that the machine, running at 600 rpm, lit about 600 lamps to an average brightness of 18 candlepower—suggesting that it could power 800 lamps at a lower, standard brightness. Francis Jehl's *Menlo Park Reminiscences* includes a particularly detailed description of this dynamo's first test, differing from Mott's in some details.[10] Both suggest, however, that, while the high-speed machine produced the power it was designed for, the difficulties of operating such a large engine at such high speeds over extended periods quickly manifested themselves. At full speed, the bearings tended to heat badly, so the machine was generally operated at under 500 rpm. According to Jehl, it continued to run satisfactorily for some weeks. In the process of shifting his operations to New York, Edison ordered a number of smaller dynamos shipped to the city and used the large machine to power the Menlo Park lighting system. Finally, toward the end of May while Edison was at Menlo Park, he ordered further experiments with the dynamo in the course of which the machine burned out due to a short circuit in the armature and was not rebuilt. The large dynamo, for all its mechanical difficulties, did what it was designed to do and proved to Edison's satisfaction that he could indeed build the giant machines he envisioned for his central station system.[11]

From the beginning of his work on the electric light, Edison made no secret of his conception of a system patterned after the gaslight. Both before and after the invention of the carbon filament lamp, he sought to learn more about gaslighting and especially about its economics. The gas service statistics collected in various cities and towns were used in planning the central station electric lighting system. This was due in part to the obvious fact that gaslight was the competition for any system of centrally distributed electric light. But gas represented more in Edison's mind than the competition—it was a guiding analogy every step of the way. The entire concept of "subdivision" of current was to a degree a product of the gas analogy, and certainly the insistence on independently controlled lights, necessitating parallel distribution circuits and high-resistance lamps, was directly modeled on a key feature of gaslighting. The model went beyond even these considerations, however, and determined other elements of the Edison system—elements that in retro-

(opposite)

Experimental Central Station in the Menlo Park Machine Shop, 1880.
The experimental station was located in the rear of the large brick machine shop. The Brown steam engine that drove the dynamos is just visible at the left.

spect were unnecessary. The best example of this extreme influence of gas is in the pursuit of underground distribution. The distribution of gas from pressurized reservoirs through substreet mains and into buildings through smaller conduits was itself modeled on urban water systems. For both gas and water such subsurface distribution seemed (and still seems) the only practical method. For electricity, however, it was an expensive and complicated alternative to overhead wires, already so familiar in the telegraph network.

No one, of course, was more familiar with the telegraph than Edison, and the possibility of overhead distribution of electricity was obvious. That large amounts of current could be carried by overhead wires had been demonstrated by existing arc lighting installations, although only for short distances and to relatively few lamps in each case. Edison's familiarity with the telegraph also included knowledge of the drawbacks of overhead wiring—frequent breakdowns due to storms, snow, and the constant exposure of primitive insulation to normal weather and city atmospheres, and the unsightliness of central districts of large cities cluttered with a maze of telegraph, fire alarm, stock ticker, and the first telephone lines. Edison was aware that one of the great attractions of gaslighting was its reliability and correctly sensed that this was due in large part to underground distribution. For electric lighting to attain a comparable reliability, Edison saw no alternative to going underground, despite the vastly greater cost and technical difficulty. There is no evidence that he was ever challenged in this logic by his colleagues, his backers, the press, or the public.

What did inhibit the creation of an underground electric system was the inadequacy of contemporary insulation. This manifested itself with frustrating persistence through 1880. The New Year's Eve demonstration had been rushed into being by the skepticism of the New York press and the anxiety of heavily committed financial backers. However, Edison never considered it a true exhibition of his system and was anxious to demonstrate the full extent of the system as soon as possible lest his vision be inadequately appreciated by the public and by capitalists. In the spring of 1880, therefore, as soon as the New Jersey clay was fully thawed, workers at Menlo Park began building the first true model of the Edison system. Once again, Mott's journal provides the best view of the frustrating job of installing an underground distribution network around the laboratory complex and, to make the model more persuasive, into the surrounding fields. On April 21 Mott recorded that "with plough and shovells the ditches for conductors to the Park Street lamps were commenced. . . ." The ditches posed few problems, so a couple of days later Mott was able to describe a device rigged up to play out the conductors as they were laid into the ditches and made the first reference to the use of tar for coating the conductor boxes "to enable them better to reject moisture and as a preventive against decay." All seemed to go smoothly, and on May 4 Mott wrote that another line was begun from the generating station. To save time and labor, Mott noted (May 5), this line and all others were to be put down in untarred boxes and covered with tar only after completely laid.

Another line was begun the following week (May 10), and Mott described the five lines that the completed system would consist of: the newly begun line of four strands of No. 10 wire to the station of the electric railway, "one of six wires North, one of 25 wires South, one of 18 wires west and one of 16 wires east." The number of wires in each main line was determined by the anticipated load: as the line made its way from the generating station, one or two wires were dropped out every hundred feet or so, so that the line tapered as it approached the end. This reduced the voltage drop along the length of the conductors by minimizing the circuit's resistance at its origin where the current was greatest. There still remained a measurable voltage drop, so the system required carefully tested and sorted lamps—those requiring higher voltages to achieve their rated candlepower were placed nearer the generators, those needing lower voltages were installed farther down the line. The Menlo Park system required consideration of such factors but was still not particularly complex.

On July 15 Mott noted in his journal:

> Conductors. The gang that have been at work on laying the conductors to the Street lamps since May 1 got them all down today, but there is still a large amount of it to tar, cover, and fill trenches. This job has taken two & one half months, with an average I should judge of six men on the work.

The next day, according to Mott, Upton began testing the insulation of those lines that had been covered. The results were that "some of the circuits are very badly insulated and all more or less defective." Mott allowed himself one of his rare commentaries:

> It seems a little strange that unexperienced men should be permitted to put down nearly five miles of wire and cover it without being required to test a single circuit or wire until the entire work is finished, and it will now require considerable extra labor and delay in putting the circuits in working order.

Mott could not have guessed the extent of the impending delay, but over the next two months numerous experiments and tests were carried out on possible insulating compounds and their behavior in the Menlo Park soil. Tests in rain-soaked ground were of special importance, and once Edison himself cabled to Grosvenor Lowrey to say that he could not leave the lab due to the fact that he had been waiting for "a wet morning."[12] Developing adequate insulation for the underground lines arose as yet another unanticipated challenge for the Edison system.

Menlo Park's most scientific minds—Upton, Jehl, and Clarke—found themselves assigned to the insulation problem, which required reliable testing of the resistance between lines. Wilson S. Howell, in charge of the installation work from the beginning, supervised efforts to develop a suitable combination of insulators. The procedure followed the familiar Menlo Park pattern of determining a promising direction, devising variations, and testing each variation under standard conditions.[13] Mott dutifully recorded the various combinations, of cloth, rubber, and wax, that Howell and his crew put to the test. As summer gave way to autumn, Edison became anxious about achieving his goal of an installation in place by year's end. When work began (for the fourth time) on laying the lines in late September, fifteen men and boys were put to the task. The final insulating compound consisted, in Mott's words, of "two thicknesses of muslin . . . with composition of Paraffine, tar, Linseed oil and Asphaltum." (According to Howell's account of many years later, as recalled by Jehl, three layers of muslin were wrapped around the wires before they were returned to the trenches.[14]) Mott reported that the first street lights were lit on Monday, November 1, and the following night "the entire line along Turnpike from Carmans to Factory was supplied with lamps and burned till nearly 12 o'clock." The story was told later that Tuesday's illumination was in celebration of the election that day of James Garfield as President. The new lines were a success, and one week later Edison's own house was put on the network. Finally, on November 20 Mott reported that the work was essentially complete, having "employed an average of about twelve men" for almost two months. The trials of the Menlo Park underground installation were a foretaste of the difficult work that lay ahead in the streets of lower Manhattan. However, Edison's success in the last weeks of 1880 convinced him and equipped him to convince others that whatever obstacles were to come could, with persistence and effort, be surmounted.

The last month of 1880 was largely given over to the job of convincing those who needed to be convinced of the completeness and usefulness of the Edison system. This had been the purpose of the Menlo Park installation, and, once the system was in place, little time was lost in exploiting it. By the beginning of December all of the elements desired for a full-scale demonstration were completed save one. Edison had very much wanted to use his large, direct-connected dynamo in a widely publicized exhibition, but difficulties in the machine's design and, more importantly, delays in the delivery of the large Porter-Allen steam engine prevented this. Consideration was given to getting a suitable engine from some other manufacturer, but Edison finally had to swallow his disappointment and plan his demonstration around the smaller long-legged Mary Ann generators that had been providing power at Menlo Park for most of the year. These were at least well-tested and reliable, and Lowrey and others from the Light Company reminded Edison that reliability was of utmost importance at this stage of wooing public acceptance. Pressures from the Light Company were building in the fall of 1880, since public opinion and financial confidence were getting restless as the first anniversary of the light's public debut approached. In early November Edison received a letter from G. W. Soren, one of Lowrey's law partners and secretary of the Light Company, advising him "of a resolution . . . to the effect that the Executive Committee is requested to visit Mr. Edison at Menlo Park early next week for the purpose of consulting with him as to measures to be taken for bringing the Company light before the public."[15] Edison may have rankled at the reference to the "Company light," but, after all, the "Company" had paid for it and, after more than two years of paying, the financiers wanted to see some returns.

To move things forward in that direction, Lowrey and others in New York City began to plan the necessary legal steps for installing a lighting system in the city. Despite all the calculations and estimates by Menlo Park workers of the size and capital requirements of such a system, Edison still seemed only dimly aware of the complexity of the task before him. In response to an overseas inquiry in October, he stated that he could "safely say that the Edison Electric Light Company of America will have one station established and in full working order lighting the lower portion of the City of New York before the first of May 1881."[16] Not only did this attitude underestimate the technical difficulties, but, as the Light Company officials tried to inform Edison, political and economic hurdles also stood in the way. After Edison sent him rough plans for laying wires in New York City streets, Tracy Edson wrote back to explain some of the steps to be taken:

> I have today called upon the Commissioner of Public Works, Allan Campbell Esq., with whom I am acquainted, and stated my desire that he should visit Menlo Park with me as soon as you are ready and see you and the Light as I saw it the other evening, in a *private* and *quiet* way, before any public exhibition or announcement is made, as he could have a better opportunity, in that way, to examine into and judge of the merits of the system.
>
> . . . Now as I think it would be a great assistance to us in getting the rights we desire if these Gentlemen should be favorably impressed in regard to our Light, I would like it very much if you would make arrangements to exhibit it to them in the manner above indicated, as soon as you conveniently can. . . .[17]

ALDERMEN AT MENLO PARK.

EDISON GIVES A SUCCESSFUL EXHIBI-TION OF HIS ELECTRIC LIGHT.

The City Fathers Partake of a Collation, Swallow Innumerable Bumpers and Make the Most Scintillating Speeches.

Late yesterday afternoon Aldermen Morris. Mc-Clave, Jacobus, Stack, Wade, Kirk, Fink and Slevin, Park Commissioners Green and Lane, Superintendent of Gas and Lamps McCormick, Excise Commissioner Mitchell and ex-Alderman Taylor visited Menlo Park to view a test of Edison's electric light. Accompanying these honorable gentlemen were G. Salyers, Paris; E. Biederman, Geneva; S. C. Wilson, and the following directors of the Edison Electric Light Company: Tracy R. Edson, Grosvenor P. Lowry, Nathan G. Miller and S. B. Eaton. Mayor Cooper was to have gone, but had to go to a fair, and consequently sent his regrets.

New York City Aldermen at Menlo Park, December 21, 1880.

This article and cartoon in the New York Truth *describe the visit by the New York City Board of Aldermen to inspect the Edison electric light system. Edison and Lowrey provided a sumptuous feast to help win approval of a plan to lay conductors for his New York City central station under the streets.*

As Edson had undoubtedly foreseen, the problem of getting the city's permission for the lighting system's installation under the streets required more than quiet demonstrations for the benefit of one or two officials directly involved; the City Fathers themselves had to be wooed.

In mid-December Edison himself was called into the act. When the President of the New York City Board of Aldermen, John S. Morris, voiced reservations about the Light Company's application for a franchise, a letter was prepared for Edison's signature that explained the care and extent of the work at Menlo Park and ended with an invitation to come out for a full-scale demonstration. A few days later, formal invitations were sent from Menlo Park to all members of the Board to come to the laboratory on Monday evening, December 20.[18] This occasion, far from the "private and quiet" showing Edson had suggested, was, in late twentieth-century jargon, a "media event" of the first order, and the newspapers made the most of it. The reporter from *Truth* found it all great fun. After describing the impression that the Menlo Park lights made on the politicians, he detailed the full extent of the lobbying effort:

> By this time the city fathers had begun to look quite dry and hungry, and as though refreshments would have looked much more palatable to them than the very scientific display they had been wondering at for two hours without a great deal of comprehension, although with a wonderful exhibition of understanding and appreciation.
>
> Their hopes were quickly realized by the announcement that the collation was ready. For half an hour only the clatter of dishes and the popping of champagne corks could be heard, and then the wine began to work and the Aldermen, true to their political instincts, began to howl, "Speech, speech." One of the witnesses of this visit said that the City Fathers were amazed at the appearance of the man they called "Professor" Edison. "Why," whispered one City Father to another, "he looks like a regular fellow. See how he handles his cigar—just like the boys in the Wigwam [Tammany Hall].[19]

More negotiations followed in the ensuing days concerning the terms under which the Edison interests would be allowed to work in the streets, but the demonstration at Menlo Park apparently left no doubt of the workability of the Edison system.

Other exhibitions through the month of December broadcast the message that the Edison light was finally ready. On December 5 Menlo Park played host to Sarah Bernhardt, the internationally celebrated French actress then in the midst of an American tour. The visit was the "Divine Sarah's" idea, for she was fascinated by the reports she had read of Edison's phonograph—an especially intriguing invention to a performer famous for her voice. Her trip to Menlo Park, however, provided yet another opportunity for putting the light before the public, and Edison did not hesitate to take advantage of it.[20] Other distinguished individuals made their way to Menlo Park that month. Lizzie Upton wrote to her sister Sadie, "The Electric Light is lovely now. I do wish you could see it. We are to have it in here tonight. All the streets are lighted and all over the fields. Jay Gould is coming out to see it tonight and Dudley Sargent is coming out here tomorrow night with his lady."[21] As in the previous year, a trip to Menlo Park was a fashionable holiday sojourn during the last week of 1880. Large numbers came to the New Jersey village on Christmas Eve, while others came to observe the second New Year's Eve display. More than 400 lamps spread through the laboratory buildings, homes, streets, and fields gave dazzling testimony to the arrival of the electrical age.

Menlo Park Journal, 1880

On Saturday, May 22, 1880, Mott began summarizing the past week's work in his journal, interspersing the summaries among the daily entries. These brief summaries, better than any other source, give a sense of the variety of activities under way at Menlo Park. They are reproduced here.

(May 22) Work general. Glass blowers pushing work on pumps. Gang laying conductors to Street lamps. Three men at old factory preparing building for lamp manufacturing. Mr. Batchelor on Machine for making small clamps, occasionally stopping to discuss gear question for Magnetic locomotive. Men changing counter shaft and pulleys on electric locomotive.

(May 29) Work general for the closing week and to day—two men preparing the sleepers for extension of R.R. Four ballasting and shapening up old track. Gang putting down conductors to street lamps[,] 2 carpenters at work in old factory. Mr. Batchelor & Force making and preparing to put up clamp machine, Cunningham & Bradley on Small Dynamos. Glass blowers on pumps. Laboratory very quiet.

(June 5) Work general. During the week Mr. Batchelor & Martin Force on clamp machine. Ott on turning Bast Fibers. Andrews on circular annealer for outside globes. Bradley and Andrews on apparatus to try the power of magnetic traction. Glass Blowers on Pumps. Two or three on extending track of electric R.R. Gang on laying conductors. Three mounting pumps. Cunningham on Small Dynamos.

(June 12) Work general. During the past week Dr. Moses continuing experiments on the reduction of Alumina and on making carbonized box wood crucibles with which to conduct the exp. but has had no further success in reducing it. Glass Blowers at work on pumps. Bradley on machine for making Bast fibers. Batchelor, Force, Ott & Andrews on different parts of clamp machine. Cunningham on small field Dynamos. Hornig on gear for elec. locomotive, Clarke on means of most economically maintaining the electro motive force in large circuits and systems. Carpenters on New cars and gang on laying conductors and a few on Rail Road extension.

(June 19) Absent fishing—Mr. Edison and about all the men except some in the shop, left here about four o'clock p.m. to take a schooner (hired for the purpose) at Woodbridge for an excursion and fish, expecting to remain away until Monday night. Work general of the week. Several in shop at work on patterns for motor gear. Men & Ayers team on Rail Road extension. Mr. E. interested and experimenting on Bast fiber lamps. Batchelor still absent. Cunningham on small field dynamos.

(June 26) Work general of week. Mr. Batchelor still absent. Ott on former for and getting out slotted plates in which to carbonize Bast fibers. Most of the men in shop on preparing the gear castings for putting on the electric locomotive. Three men during day on R.R. extension and Martey [?] with five men from 6 o'clock till dark. Mr. Edison and Upton testing and experimenting with Bast fiber lamps. Glass blower on pumps.

(July 3) Work general of week. Mr. Batchelor, Ott & Force on clamp machine. Mr. Edison & Upton, testing lamps and experimenting on carbons. Pump power put up in lamp Factory. Men on R.R. extension, several in Shop on Gear of electric locomotive. Glass blowers on pumps. Carpenters on Cars & elevated messenger exp.

(July 10) Work general for past week. Mr. Batchelor, Force and Ott principally on clamp machine, the former a little on captive baloon [balloon]. Smith and one or two assistants on Magnetic brake. Logan and Martin on gear of electric locomotive. Several at Lamp Factory. Gang on R.R. and on Conductors. Dean finishing up wood miller. Glass blowers on 5th hundred pumps. Bradley, Flammer & Van Cleve on experimental carbons. Upton & Clarke still absent.

(July 17) Work general for past week. Men at work at Gas and Air works and fixtures at Lamp Factory. Men finished the preliminary laying of the conductors to Street Lamps. Dean finished wood miller, took it apart and put away. Gear of electric locomotive about finished. Mr. Batchelor again interesting himself in experimental carbonization. Messrs. Edison and Upton experimenting with and testing lamps.

(July 24) Work general. Men preparing the gas carbonizing furnaces by putting in the gas and blast pipes and fixtures. Clarke on Electric Locomotives. Mr. Batchelor on carbons and apparatus for carbonizing. Conductor gang uncovering street lamp circuits. Men finishing up gear for electric locomotive.

(July 31) Work general for past week. Messrs. Edison and Batchelor making some interesting experiments with carbons and hydrocarbons and means of carbonizing. Andrews making carbon formers of different devices for us in large 30 carbon moulds of gas furnace. Dean making fiber clamps and formers for four, six, and eight inch loops. Mr. Clarke on electric locomotive, insulated car wheels and roller creepers [?]. Wilber here three or four days drawing specifications. Cunningham making glass heater or the device for holding tubes for heating over gas flames (described under date May 28). Several tables fitted for glass blowers at Lamp factory. Hammer making the rubber connections between pumps and mercury pipes. Conductor gang wrapping and tarring the 25 wire circuit. Carpenters building the large magnetic separator.

(Aug. 7) Work general for past week. Team drawing coal dust from ten cars of which are laying on the switch for Mr. Edison. Men continuing work cleaning out the Pond. Ott remoddling and putting screw gear in telephone receivers. Dean making former in which to cut Bamboo fibers. Mr. Batchelor carbonizing and experimenting on apparatus for that purpose and has worked it down "fine". Mr. Wilber here most of the week getting up

specifications. Men preparing lamp factory for occupancy by putting in the final trimmings. Chalk motograph receiver tried O.K. to noisy & armature to be constructed differently.

(Aug. 14) Work general for past week. Mr. Batchelor experimenting to determine the cause and devise means to prevent the long carbons from bending over side-ways in the lamps, found the cause and remedied the defect by removing the cause. Men under Logan putting the cast gear on the electric locomotive. Trying the experiments with the coal dust furnace and experienced great delay and annoyance by it. N.G. Testing the blower and Gas supply at Lamp Factory, the former efficient but the latter has neither capacity nor simplicity for practical use there. Men wrapping conductors with muslin tarred and then wound with marlin and again tarred. Mr. Wilber here several days at work on Patent Specifications.

(Aug. 21) Work general for past week. Men putting the gear in the electric locomotive. Dean making cutting clamps & former for Bamboo strips. Andrews making carbon formers. Ott & McKenzie on call box for Pond indicator. Upton making calculations of the power, conductors, &c. necessary for a lighting station in New York. Clarke working on Dynamo [illegible] Book No. 116. Mr.

Batchelor superintending starting the Lamp Factory.

(Aug. 28) Work general of past week. Mr. Batchelor on the drawings and details of clamp machine. Dean with several assistants at work on same and on Bamboo cutters & splitters. Upton and Hammer on size, length, &c. of conductors for lighting station in New York. Clarke still on New form of armature adapted for large machines. Smith working to improve the clutch of friction wheel of electric locomotive.

(Sept. 4) Work general. Mr. Batchelor on detailed drawings of clamp Machine and Dean and assistants at work on same and on clamp former for Bamboo. Smith remoddling the friction clutch wheel for electric locomotive. Men digging foundations &c. for building and gas works at Lamp Factory and pumps there prepared for starting.

(Sept. 11) Work general of past week. Mr. Batchelor making efforts to get the Factory started. Messrs. Edison, Upton and Hammer at work on conductors for central Station and plans for laying them. John Ott making McKenzie call boxes for Pond indicator. Dean working several men on Fiber cutting moulds and on clamp machine. Carpenters splitting old glass house for Draughting room. Other carpenters at work on preliminary work of

Supplemental Factory building. Johnson commenced on Chandelier.

(Sept. 18) Work general past week. Experiments continued in insulating compounds. Mr. Batchelor working at Power Mercury Pump at Factory, but unsuccessful in getting it to work reliably. Mr. Upton at work on Station conductors. Smith on wooden facsimile of large armature commutator connections. Logan on casting of Large Dynamo. Dean and assistants on Clamp Machine and fiber formers.

(Sept. 25) Work general past week. Carpenters at work on building supplemental to Lamp Factory. Gang commenced third laying of Street Lamp conductors, insulating with composition. Large planer running on castings etc. of Large Dynamo. John Ott finished three call boxes for Pond indicator and continued work on wooden facsimile of commutator for Large Armature. Dean on Clamp Machine and Bamboo clamp formers.

(Oct. 2) Work general for past week. Logan & some of shop men working on Large Dynamo. Dean and assistants on Bamboo clamps and on clamp machine. Painters finishing up new drawing office. Factory turning out quite a number of lamps and Francis testing them. Upton and Hammer on Station conductors. Mr.

Batchelor at Factory. Johnson at work on chandeliers.

(Oct. 9) Work General for past week. Factory has been turning out and Francis testing a considerable number of lamps. Work progressing nicely on the large dynamo. Dean and assistants still on Clamp machine and fiber cutting clamps. Gang on insulating the Street conductors. Carpenters on inside work of Factory supplemental building. Man. (Moore) left for China and Japan.

(Oct. 16) Work general. Testing lamps by the quantity. Men on Large Dynamo. Gang of about twelve on Street conductors. Dean and assistants on clamp Machine and clamps for cutting bamboo. Rotary engine received and partially tried.

(Oct. 23) Work general for past week. Carpenters finished Black [illegible] Shop. Dean made a start on the large armature. Mr. Batchelor experimenting on pumps. Gang on insulating cables. John Ott at work on Meters. Draughtsmen all in new room.

(Oct. 30) Work general past week experiments conducted on heating carbons in gasses. Dean with assistants working on large armature and on mould for cutting bast fiber. Gang still on insulating conductors. John Ott finished wheel meter. Gas fitters running steam piping at factory for heating purposes. Mr. Batchelor experimenting on and having pumps changed to but one tube with dryer and spark gauge attachment. Men at work preparing to pump water from gulley.

(Nov. 6) Work general for past week. Upton & Hammer tabulating the statistics obtained by Mr. Russell's canvass. Edison & Francis treating carbons in gasses. Commenced putting out the Street lamps. Dean & assistants on Armature. Cunningham on commutators for same. Ott on Meters. Several men in shop on Magnets &c of large dynamo.

(Nov. 13) Work general of past week. Messrs. Batchelor & Edison engaged principally in Interference case. Little doing at Factory in consequence of delay in getting the screw pump. Men in shop pushing work on large dynamo & armature. Ott at work on an Electric dynamometer.

(Nov. 20) Work general past week. Bed plate for Dynamo Engine placed in shop and some preliminary fitting of the parts effected. Discs secured on the Armature shaft, and work on the commutator &c. progressing satisfactorily under Dean. Messrs. Edison & Batchelor experimenting at factory on times for subjecting fibers to process of carbonization, on length of time and heat to apply to lamp on pumps, and on heating mercury and pumps.

Mr. Clarke working on the relation and laws of motors to machine &c. Acheson on [illegible] carbon by Electrolysis.

(Nov. 27) Work general of past week. Mr. Batchelor and assistants preparing the pumps and pipes for the power pump and clean mercury. Mr. Edison with two or three assistants making experiments on high vacuum on pumps in Laboratory. Dean and several assistants pushing work on the large armature. Logan and other on Magnets, base, etc. of large dynamo.

(Dec. 4) Work general past week. Work on large dynamo, Glass blowers making pumps. Mr. Batchelor preparing for power pump. Mr. Edison experimenting on high vac[uum] and lamps.

(Dec. 11) Work general of past week. Screw Mercury pump put up and prepared ready for work. Work in shop progressing on large dynamo. Logan winding cores of magnets. John Ott on a meter with four Mercury cups for contacts. McKenzie trying to work call box with magneto current generated by the magneto call box of Johnson. Mr. Clarke making measurements to determine the loss in dynamo by local cutting. Nichols determining the constants for the low resistance Electric dynamometer.

(Dec. 18) Work general for the past week. In the shop work

has been pushed forward on the large dynamo, and Ott finishing up lamp sockets. At the Factory the glass blowers on pumps and lamps, and Moffett mounting and putting up new device for securing the pumps adopted, and the castings ordered. Lawson working on plating the ends of carbons and Herrig experimenting on plating carbon and wires together. Second experiment on burning dust coal direct under boilers tried and so far with fair success. Lamps being tested and put on the lines preparatory to illumination on Sunday and Monday nights.

[Mott returned from a four-day Christmas vacation on Tuesday, December 28 and recorded the following report of the Christmas Eve demonstration].

(Dec. 28) 400 lamp test. 400 lamps were burned on Friday evening, Christmas eve, and Economical test made of the Engine by Mr. Clarke, boilers forced to 90 lbs. pressure, engine to 75 Rev., gave 6.9 lamps per horse power, but I believe this result is obtained after deducting the power consumed in friction of shafting etc.

(Jan. 1, 1881) Illumination & Economic test. In the evening 408 lamps were burned and a number of visitors were here to see the display. Mr. Clarke made a test of the engine while running for the lamps and developing 82.3 horse power and got perfect diagrams. Boiler pressure 110 lbs. Revo 75, total H.P. 82.3 less frictional diagram left net of 61.95 H.P. or on 408 lamps gave 6.59 per H.P. less field gave net 7.88 per h.p. on 22.58 pounds steam per H.P. per hour. Proving the Brown engine first class for economy etc.

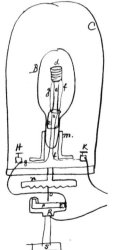

CHAPTER 7

Promises Fulfilled

In 1880 Edison's establishment at Menlo Park changed its character from a laboratory primarily devoted to invention to a site for the development and manufacture of components of the Edison lighting system. The little New Jersey village was not well suited for the more mundane role of manufacturing center, especially since most of its products were intended for the system's great inaugural installation in New York City some twenty-five miles away. The abandonment of Menlo Park, therefore, began as soon as the demonstration system had served its purpose. The year 1881 saw the gradual diminution of the laboratory and the scattering of much of the crew of assistants who had been so much a part of Edison's work there.

Charles Mott's daily records of the laboratory's activity reflected these changes. Throughout January Mott generally included a reference to Edison's presence in New York, frequently adding that others, such as Clarke or Kruesi, were accompanying him. In February the move accelerated: Charles Batchelor sailed for Europe to arrange for electrical exhibitions in Paris and London, and Edison settled on a permanent location for his New York offices at 65 Fifth Avenue.[1] On February 7 Mott noted that his brother Sam had received orders to move his drafting operations to New York, and a month later, on March 10, Mott wrote that he was "Ordered by Mr. E. to go to N.Y. office with Elec. Light Patents & caveats, so my records must cease from this date."[2] (Mott, in fact, continued to keep a daily record in New York for some time.)

As the electric light moved from the laboratory into the working world, and as Edison moved himself from Menlo Park to New York, the kind of records that chronicle their story changed. However eccentric the Menlo Park laboratory mode of operation, it followed a basic rule that Edison had learned early in his career—whatever happens, write it down. Despite problems in dating, disjointed sequences of notes, haphazard organization, and cryptic references, the laboratory notebooks still stand as a remarkably full account of activity. The concentration of work in one or two buildings, and the close supervision such concentration allowed, made it possible, despite the often hectic atmosphere of Menlo Park, to maintain records of such obvious value (although imperfect) that they were carefully kept, with little attrition, until they passed into the hands of archivists. Edison's New York years (1881 to 1886, when he moved to West Orange, New Jersey) produced a written legacy of a very different quality, characterized much more by the demands of business and bureaucracy than by the procedures of a laboratory and the possibilities of patents. The kind of insight the notebooks afford into the inner workings of the laboratory and the thinking of the men who worked there is not available once the scene shifts to New York. The records from 65 Fifth Avenue and the half-dozen other centers of Edison enterprise in the metropolis are largely the correspondence of business and finance and those technical notes deemed significant enough to keep. The story of Edison in New York is therefore different in both tone and substance from that of the Wizard of Menlo Park.

The ground for the move in early 1881 was well prepared. The most important step was the incorporation on December 17, 1880, of the Edison Electric Illuminating Company of New York by some of the same investors who had formed the Edison Electric Light Company two years before. Like the 1878 incorporation, this one was organized by the indefatigable Grosvenor Lowrey, who still saw himself as a protector of Edison's personal interests in the ever more complex financial and political dealings surrounding the Edison enterprises. As a matter of fact, Lowrey's loyalties were already shifting more and more to the Wall Street interests he had served so long, but this was perhaps not yet obvious to Edison. On the date of the Illuminating Company's incorporation, Lowrey wrote to report the event to Edison and to soothe, once again, the inventor's prickly ego:

My dear Edison:
I shall not present your letter of resignation as Mr. Fabbri very strongly objects to your leaving the Board. His impression was that "Edison's name is a tower of strength to us, and if he never attended a meeting, it would be a great loss if his name should not appear at all times among the names of the Directors." . . . We yesterday organized and filed the articles of association of the "Edison Electric Illuminating Company of New York," under the general gas company acts of this state, stating the object of the organization to be to illuminate the streets, &c. by gas. We have to state this as the object in order to perfect a legal incorporation, but every gas company has by law, after it is organized, the right to turn itself into an Electric Light Company, and we have prepared a long ordinance to be submitted to the Common Council if we are so advised, granting us the franchise to lay down wires over the entire city.[3]

The formation of the Illuminating Company was part of Lowrey's carefully orchestrated plan to clear away all conceivable legal obstacles to the installation of the Edison system in New York. As was explained in the 1881 annual report of the Light Company, the new organization was required by New York State law restricting use of the streets for lighting distribution to firms incorporated under the Gas Statutes, as opposed to the general corporation laws. The legal necessity to follow the gas industry pattern in the formative years of electric lighting accounts in part for the separation in America between electric utilities and electrical manufacturers (as distinguished from the pattern that developed in the communications industry, for example). In real terms of finance and management, however, the separation was (as in the case of many early electric utilities) initially slight, for of the thirteen members of the Illuminating Company's first Board of Directors, nine were from the Light Company's Board. The first president, Norvin Green, was also president of Western Union, and the Drexel, Morgan interests were well represented, most visibly by Egisto Fabbri, who served as first treasurer of the company.

The visit of New York's Board of Aldermen to Menlo Park followed the incorporation by only three days, as Lowrey lost no time in pursuing the required franchise for the Illuminating Company. Lowrey's haste was caused not only by his impatience to advance the light to a moneymaking stage but also by emerging competition. On December 18, the first public electric lighting system in New York City went into operation when the Brush Electric Light Company started up its three dynamos at 133–135 West 25th Street to supply arc lights on Broadway between 14th and 34th Streets. While a year ago the New York newspapers had been filled with news of the miracle wrought with carbonized thread at Menlo Park, this December they devoted their fanfare to the spectacle of the glaringly bright Brush arc lamps that were giving birth to Broadway's "Great White Way." The Brush system used the best arc lighting technology of the time, generators of Brush's own design (each sufficient to power sixteen 2,000-candlepower lamps), and overhead transmission lines on telegraph poles. The Brush Company operated the Broadway lamps as a free demonstration for almost six months, after which it (now the Brush Electric Illuminating Company) was rewarded with a city contract to light not only the original Broadway stretch but also nearby portions of 14th Street, 34th Street, and Fifth Avenue, as well as Union and Madison Squares. By the end of 1881 there were fifty-five Brush arc lamps as part of New York's public lighting system. In addition, the Brush light quickly made its way into nearby hotels and theaters in what was then the established shopping and entertainment district of the city. While the Brush arc light represented, to Edison at least, no major technological achievement and no direct challenge to the incandescent light, it was highly visible and received ready public acceptance at precisely the time that people were growing openly weary of the failure of the Edison

system to make its long-heralded commercial appearance. This surely must have given Edison a special incentive to move further activity from the laboratory into the streets.[4]

In the same article of December 21, 1880 in which it reported the aldermen's visit to Edison's laboratory and the new arc lights on Broadway, the *New York Post* commented, "There are now six different companies at work introducing electric lights in this city, the lights being known as the Brush, Maxim, Edison, Jablochhoff, Sawyer and Fuller (Gramme patents) lights." While most of the activity of Edison's rivals was limited to arc lighting, the public did not always discriminate in its perception of electric lighting, which accounted somewhat for the diminished spotlight on Edison's efforts during the busy New York years. In addition, the competition was not exclusively from arc lighting. The same *Post* article described the success of Hiram Maxim's incandescent lights "in the vaults and reading rooms of the Mercantile Safe Deposit Company," where they had been "working admirably" for almost two months. At this time there were still no Edison installations outside of Menlo Park (except for the steamer *Columbia*). Edison had no doubt in his mind that the Maxim lamp was nothing more than a bald copy of his own cardboard filament

Brush Arc Light, 1880.
At the end of 1880, the Brush arc light lit Broadway's "Great White Way," and a year later there were fifty-five Brush lamps in New York's public lighting system. The light also quickly made its way into hotels and theaters. While the Brush arc light did not directly challenge Edison's incandescent light, its high visibility and ready public acceptance, combined with the failure of the Edison light to make its long-heralded commercial appearance, provided a special incentive to move activity from the laboratory to the streets in preparation for the installation of the Pearl Street central station. (Courtesy Smithsonian Institution)

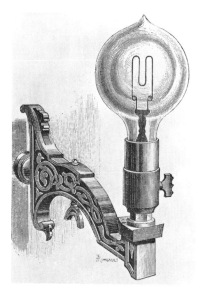

Maxim Lamp, 1880.
*While most commercial rivals pro-
moted arc lighting, some competitors
began to introduce incandescent
lamps. Hiram Maxim's unique M-
shaped filament was only an outright
copy of Edison's own cardboard filament lamp, but installation of the
Maxim light in the vaults and reading rooms of the Mercantile Safe
Deposit Company served notice to
Edison and the Electric Light Company that they could wait no longer to
commercialize their product. (Courtesy Smithsonian Institution)*

lamp, distinguished only by its M-shaped filament. Nonetheless, Maxim's
success in the Equitable Building (where the Mercantile's offices were
located) was a worthy technical achievement and served notice on Edison and the Electric Light Company that they could wait no longer to
commercialize their product.[5]

Lowrey's push to organize the Illuminating Company and clear away
the legal obstacles to the New York central station was one response to
the new pressures to get the system into operation. An even broader
response was Edison's own organizing of the manufacturing capability
that the central station system would require. The first part of this effort
had been the establishment of the lamp factory at Menlo Park, which by
the end of 1880 was turning out several hundred bulbs daily. The lamp
factory set a precedent for the organization of manufacturing for the Edison system, for it was independent of the Edison Electric Light Company except for a license to manufacture and supply lamps under the Edison patents. The financiers behind the original company had no intention
of putting more of their money into the electric light before they could
realize some return on the patents for which they had already supported
Edison for more than two years. Edison, for his part, was apparently
quite willing to put his own money into an establishment over which he
could expect to retain control. The Light Company, while happy to see
Edison's money financing lamp production, was careful not to give too
much away. When, for example, a contract was drafted in January 1881
between it and the "Edison Electric Lamp Company," the Light Company directors not only balked at the five-cent-per-lamp profit that was
offered, but also objected to the proposed company, to be made up of
Edison, Batchelor, Upton, and E. H. Johnson. The contract that eventually emerged in March was between the Light Company and Thomas
Edison alone and reduced the allowed undivided profit per lamp to three
cents, after which the thirty-five cents the Company paid for each lamp
was to be cut by half the excess profit.[6] This episode illustrated the extent to which the Light Company managed to exercise ultimate control
over the Edison light, even in opposition to the inventor himself.

Another pattern that emerged with the organization of the lamp works
was the disbanding of the once close-knit crew that had surrounded Edison at Menlo Park. As the new manufacturing concerns formed in 1881,
men like Upton, Kruesi, Dean, and others found themselves in management roles, often at some distance from Edison himself. The lamp factory was at first closely supervised from the laboratory, with Batchelor
taking a leading hand in getting everything under way. It began turning
out bulbs in midsummer 1880, but the official start of factory production
was designated as October 1, 1880, and the first regular payroll began on
November 11. When it was decided in January 1881 to send Batchelor to
Europe to organize the electric light interests there, Upton was placed in
charge of the lamp works. This was to mark a permanent change in Up-

ton's career; for the rest of his working life he was primarily concerned with light bulb production. He never again worked closely at Edison's side, except for a short period in late 1881 when Edison went to Menlo Park because (as Edison wrote Johnson) "Upton had got away off his base & was trying to get back without informing me."[7] By early 1881, in fact, Charles Clarke had become the one Edison leaned on for the kind of mathematical and technical assistance he had once sought from Upton.

Other key Menlo Park figures soon found themselves scattered among the proliferating Edison enterprises. John Kruesi was put in charge of the Edison Electric Tube Company, set up at 65 Washington Street in New York in February 1881 to organize the installation of the underground distribution system for the central station. Charles Dean, originally Kruesi's assistant in the Menlo Park machine shop, was soon afterwards designated superintendent of the Edison Machine Works, located in the former Aetna Iron Works at 104 Goerck Street on New York's lower East Side. These three enterprises, the lamp works, the Tube Company, and the Machine Works, became the manufacturing arms of the Edison system and formed the foundations of the system's corporate culmination, first as elements of the Edison General Electric Company, organized by Henry Villard in 1889, and then within the even larger General Electric Company, created by merger with Thomson-Houston interests in 1892.

It was with the immediate future in mind, however, that Edison began organizing the manufacturing companies in the winter of 1881. Lamps were the most novel of the system components, and placing their manufacture on a sound commercial footing presented the most obvious challenge. This can be seen as one of the reasons for the close link, both geographic and administrative, between the lamp works and the Menlo Park laboratory. With lamp production under control by the end of 1880, it was time to develop the means for supplying the remaining elements of the system. These, however, required less radical departures from ordinary machine shop practice, hence the relative ease with which the two chief mechanics from Menlo Park were able to establish their New York works. Charles Mott's daily record of Menlo Park activity noted Kruesi's absence (in the company of Edison and others) in the first part of February 1881 when, presumably, a search was being made for a suitable New York building for the Tube Works. The Works formally began operation on February 14,[8] but Mott noted on the 18th that Kruesi was at the laboratory "making exp. with "insulating" compound No. 7 varying proportions of Parafin, preparatory to commencing operation at building in Washington St."[9] The Menlo Park lab served all of the New York shops, as well as the lamp works, as an experiment and testing center, more like a twentieth-century industrial laboratory than has been generally recognized. The new factories, however, soon developed testing facilities of their own, and gradually, during 1881, even the more experimental work, such as the search for a better meter, was transferred to New York.

Kruesi worked closely with Edison in managing the Tube Works, striving to quickly organize the creation of an underground distribution network. The model underground system at Menlo Park had taught both of them much about the requirements for a reliable installation. The insulating compound of asphalt, linseed oil, paraffin, and beeswax that Wilson Howell had so laboriously developed was adopted with little change. Kruesi soon devised a standard two-wire conduit that consisted of two copper conductors, with semicircular cross sections, separated by appropriately shaped pasteboard washers and covered by insulating compound inside a protective iron pipe. Further work was required to develop various junction boxes, fuse holders, and other elements of a full-scale underground installation. In addition, the performance of the conduits and connections had to be constantly tested as they were put down under New York streets—another valuable lesson from the Menlo Park experience. All of this required close coordination between the Tube Works and Edison's headquarters on Fifth Avenue, and Edison and Clarke spent many hours either in the shops or in the streets with Kruesi, solving problems caused by unreliable suppliers, careless laborers, or unpleasant surprises like one reported to Edison by Kruesi in June 1882:

> We have observed a very unexpected phenomena in our Junction Safety Catch boxes. Friday night we opened one which was put in long ago, when the inside cover was taken off a lot of gas escaped, and when one of the men came near with a light it blazed up and burned for about 5 minutes (singed a man's hair) and was put out with a pair of bellows. Today again we had to open one in daytime—the escape of gas was the same only was not ignited.
>
> It may cause trouble if a man had to breake a connection when the current was on—the spark may ignite it. I suppose it comes from the compound. We may have to put two plugs in the inside cover & blow out the gas occasionaly—what do you think of it? [10]

The solution of problems of this sort became almost routine, and none presented long-term difficulties. Indeed, the basic underground distribution technology was spelled out by Edison in a patent filed April 22, 1881 (U.S. Patent 251,552, issued December 27, 1881), just as the work in the New York streets was getting under way, and only one notable change was made before the two-wire system was replaced by three-wire distribution in late 1882. This change was the replacement of the cardboard spacers in the conduits by ordinary rope, which could be easily wound around the conductors to provide the needed separation. In terms of sheer toil, nothing in the building of Edison's "First District"

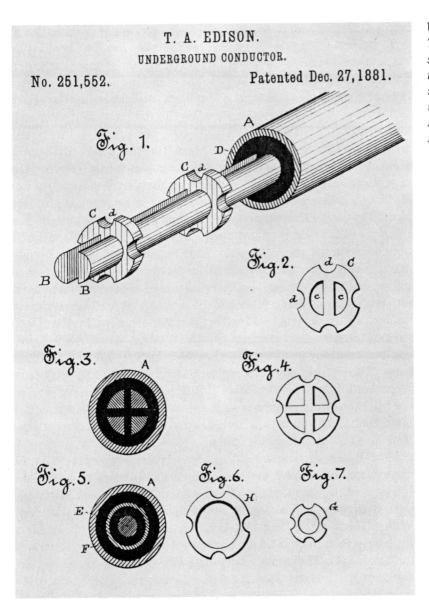

T. A. EDISON.
UNDERGROUND CONDUCTOR.
No. 251,552.   Patented Dec. 27, 1881.

Underground Conductors.
This patent drawing illustrates the standard two-wire conduit used in the Pearl Street district. They were made at the Electric Tube Works in New York under the direction of Edison's master mechanic, John Kruesi.

matched the effort of Kruesi and his Electric Tube Works, and the former Swiss mechanic was given ample credit for his labors.

Construction of the conduits and junction boxes was not the only technical challenge presented by the distribution system. The fundamental problem of transmission losses through long lengths of conductors had long been a concern of Edison, and the New York system was the first time that a solution was critical to success. In January 1880, a year before moving to New York, Edison filed a patent application for a system of "multiple-arc distribution," which was the most fundamental statement of the principles of power distribution on a parallel circuit. Indeed, the

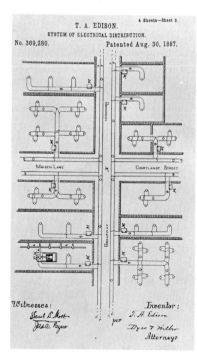

Proposed Electrical Distribution for New York City, 1880.
Edison filed his patent application in January 1880. Its description of "multiple arc distribution" was his most fundamental statement of the principles of delivering power through a parallel circuit. Indeed, the claims were so broad that the patent was not actually granted until 1887, by which time its claims were rapidly being superseded by the advancing state of the electrical art.

claims were so broad, the patent was not granted until 1887 (U.S. Patent 369,280), by which time its features were rapidly being superseded by the advancing state of the electrical art. Of greater practical value was Edison's invention in the summer of 1880 of the "feeder-and-main" principle of distribution, which he covered by several patents (especially U.S. Patents 239,147 and 264,642). Without the ingenious feeder-and-main design, the Edison system might have been stopped dead by the enormous cost of the copper required by a "tree" circuit like that used at Menlo Park. In tree-type distribution the drop in voltage between the generator and the farthest end of the circuit had been reduced by the use of thicker, lower-resistance conductors near the generators, gradually tapered to thinner diameters toward the outer limits of the system. For the New York system, with service extending as much as a half-mile from the central station, this would have required enormous and costly mains. The feeder system bypassed this problem by supplying power to the service mains through a number of smaller conductors—the feeders—each serving only a portion of the mains. A relatively large drop in voltage, say ten percent could be tolerated in the feeders, for the drop in the local circuits—service mains and house wiring—would be negligible and thus relatively unaffected by changes in load. The feeder-and-main network was simple as well as vital to the success of the Edison system. When Lord Kelvin, the renowned British physicist, was asked why no one had ever before thought of such a straightforward solution to such a fundamental problem, he was quoted as replying, "The only answer I can think of is that no one else is Edison."[11]

Most of Edison's time in New York was divided between work on the underground distribution system, under the auspices of Kruesi's Tube Works, and the development of key system elements—primarily the large dynamos—in the Goerck Street shops of the Edison Machine Works. The Machine Works were initially a much expanded version of the Menlo Park machine shop, but the combination of specialization in the construction of dynamos and the steadily expanding commercial demand for its products soon changed its character. The move to 104 Goerck Street, supervised by Charles Dean, began on March 2 after arrangements for the use of the buildings was completed with the shipbuilder John Roach. Like the other new Edison enterprises, the Machine Works belonged to Edison and his colleagues and operated on a license from the Edison Electric Light Company. The scale of the Machine Works, however, quickly outstripped all the others; less than 18 months after the plant began operations, the *New York Tribune* published this glowing report:

> The manufacture of the dynamo machines and engines for working the same is controlled by a company having works at No. 104 Goerck-St., New-York, and the energy with which this part of the

enterprise is conducted may be appreciated when it is stated that within three weeks after the shops were rented the first Edison steam dynamo, capable of supplying over 1,200 sixteen-candle lamps, was tested. Until recently 300 skilled mechanics were employed, but the business has so increased that 500 additional men have been taken on, thus increasing the payroll to 800. Still more men are needed, but it has been found a difficult task to select competent men for the work. Not less than 1,000 men have been specially trained at these works since the start, a special and careful training being of absolute necessity. Since the time of testing the first steam dynamo, eight others of still larger size (1,400 lights) have been completed, and work on twelve others of the same size commenced, and during the last twelve months 355 dynamo machines of lesser capacity, varying from 15 to 250 lights, but chiefly for 60 full lights each, aggregating 75,000 lights in all, have been turned out.[12]

The Machine Works were expanded not so much to fill the needs of the New York system as to supply the many smaller installations Edison began selling in 1881. The large capacity of the Goerck Street factory, along with that of the Lamp Works, enabled the Edison interests in 1881 to respond at last to the demand for small systems that had been put off for so long.

Ever since the announcement of the carbon lamp breakthrough in late 1879, Edison had consistently turned aside the many requests he received for the immediate installation of his light in factories, stores, public buildings, streets, and the like. The one "isolated" system that he installed in 1880 was on the *Columbia*, and that was a special favor to Henry Villard. If other backers had displayed Villard's enthusiasm and eagerness, Edison might have found it difficult to refuse to install more small systems, but they did not, and Edison found it easy to put off others with the explanation that his central station system took priority. By the beginning of 1881, however, his resistance to isolated plant business diminished, and the building of Edison plants gained an enormous momentum in the following months. The change in Edison's attitude can be seen in his reply to an inquiry from Owen A. Gill, who was interested in lighting the Maryland State Penitentiary in Baltimore:

Menlo Park, Jany. 29, 1881

You perhaps know that all my efforts have been and all my appliances are devised especially for the general distribution of electricity throughout a city to be sold by meter, and not for the lighting up of a single building hence I am at the present moment at a slight disadvantage when asked to light up a single building. I could very

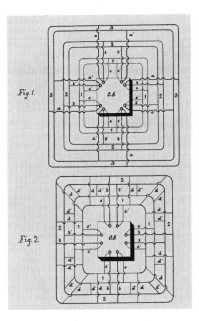

Feeder and Main Distribution System, 1881 Patent.
The establishment of a distribution network in New York was the occasion on which transmission loss through long lengths of conductors first became a practical problem for Edison. Several patents covered Edison's solution, the "feeder and main" system that supplied power to the service mains through a number of "feeders," or smaller conductors, each serving only a portion of the mains.

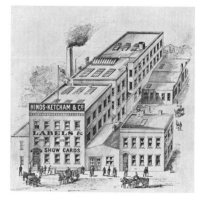

Hinds, Ketcham & Company, c. 1881.

The lithography shop of Hinds, Ketcham & Company was one of the first factories lighted by an Edison isolated plant. Electric lighting allowed the shop to operate at night without difficulty in distinguishing colors. (Courtesy Smithsonian Institution)

much easier light up a square mile with 1500 to 2000 houses than I could a single building although that may seem a paradox to you.

We are getting our offices in New York and I expect very soon to accomodate my system for isolated lighting. . . .[13]

The next week, Edison was situated in offices at 65 Fifth Avenue, and one of the first accomplishments in New York was the installation of a lighting plant at the lithography shop of Hinds, Ketcham & Company at 213 Water Street. As in the case of many of the early isolated plants, the choice of customer was not entirely arbitrary, as evidenced by the letter Edison sent to the paintmaker Louis Prang on February 11: "Last week we lighted up the Lithograph establishment of Messrs. Hinds and Ketcham in New York. They were unable to work at night until they put in the electric light. Now I learn they have no difficulty in distinguishing colors."[14] In the following months, lighting plants were installed in steamships, hotels, railroad shops, newspaper offices, and a variety of mills and factories.

For almost a year this was done largely under the direction of the Electric Light Company itself until, in November 1881, the Edison Company for Isolated Lighting was organized by some of the Light Company officials, who managed the new firm from their Fifth Avenue headquarters.[15] By the time the dynamos were turned on in New York's First District, in September 1882, there were 99 isolated plants installed across the United States and many others overseas. They ranged in size from two installations in cotton mills in Fall River, Massachusetts, that each lit 750 16-candlepower lamps, to a few miniature installations powering only fifteen lamps in, for example, Henry Draper's laboratory at New York University and part of George Eastman's photographic supply house in Rochester.[16] The isolated plants were not cheap—Edison quoted to a steamship owner a price of $2,800 simply to install generators and auxiliary equipment[17]—but the Isolated Lighting Company rapidly identified specialized markets for which the new technology had particular attractions.

Fully one-quarter of the first hundred clients were textile mills, where the superiority of electric lighting over gas in such fire-prone places was readily apparent. One mill owner in Newburg, New York, for example, wrote that "I expect the difference in insurance rates will pay the whole expense inside of two years."[18] The Edison Electric Light Company regularly put stories of fires caused by gas systems in its *Bulletins,* and salesmen were encouraged to make use of them and to explain the virtues of Edison's "safety catch" (fuse) in practically eliminating the potential for fire with electricity. The effectiveness of this appeal is apparent in the adoption of isolated lighting plants for sugar refineries, newspapers and other printing establishments, and dry goods manufacturers.

The rapid growth of the isolated plants business provided important support to the Lamp and Machine Works, which for years depended on isolated systems for much of their market. The proliferating small installations also fostered the growth of the third manufacturing establishment serving the Edison system, that of S. Bergmann at 108–118 Wooster Street in lower Manhattan. The German-born mechanic, Johann Sigmund Bergmann, had worked with Edison for six years in Newark before leaving to set up his Wooster Street shops for the manufacture of "Hotel and House Annunciators and Electrical Apparatus of all Kinds" (according to his letterhead). Due to Edison's high regard for Bergmann's work and his reluctance for most of the Menlo Park years to get too deeply into manufacturing, he had turned to Bergmann from time to time to produce his inventions—the phonograph being the most notable example. By early 1881, Bergmann's shops had expanded considerably and had just the capability Edison required for the production of the smaller elements of the lighting system. Therefore, in April Edison approached Bergmann with a proposition—he would provide Bergmann with additional capital for expansion, and would endeavor to devise appropriate products for him to make, if Bergmann and E. H. Johnson (who had earlier bought into the Bergmann firm) would agree to form a partnership devoted to the manufacture of "special appliances connected with Electric Lighting."[19] Examples of such appliances were "small fixtures, lamp sockets, switches, safety catches, shades, etc., etc." Bergmann agreed, and the Wooster Street factory became "Messrs. Bergmann & Company," which over the following years developed most of the visible consumer goods that accompanied the spread of electric lighting. Bergmann & Company, like the other Edison firms, expanded rapidly and moved into larger quarters (buildings formerly used by the Maxim lighting company) at 17th Street and Avenue B on Manhattan's lower East Side in the summer of 1881. In 1889 it became part of the newly formed Edison General Electric Company and thus provided the foundation for the enormous electrical consumer products industry.

The commercialization of the Edison system through isolated lighting plants provided not only needed markets for the new manufacturers but also an important stimulus to broader innovation. Technology developed for central stations could not be transferred to smaller installations unchanged, so the work on problems of the urban central station was accompanied by important modifications for the wide range of small systems. The most obvious difference between the central station and the isolated system was in the size of the power plant. Edison was unwavering in his belief that central stations would require large direct-connected dynamos of the type that took so much time and effort to build at Menlo Park in 1880. He saw the primary products of the Goerck Street works as large dynamos rated at 100 horsepower or more and capable of pow-

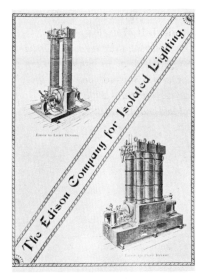

Catalog for the Edison Isolated Lighting System, 1882.
Officials of the Electric Light Company, eager for a more immediate return on their investment than the central station system promised to deliver, established the Edison Company for Isolated Lighting to market isolated systems. Catalogs and brochures promoted the single-plant installations.

Electric Lights for the Vanderbilt Mansion, 1881.
This notebook drawing was a preparatory outline of the lighting for the Vanderbilt mansion, one of the first private residences with an Edison isolated plant.

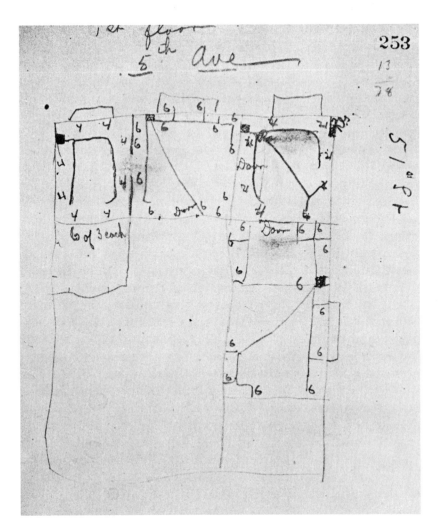

ering as many as 1,200 full-size lamps. Isolated plants, however, would not require such large, expensive machines, so facilities were needed to produce smaller dynamos.

There existed, of course, a well-tested and reliable small dynamo design very familiar to Kruesi and his associates at the Machine Works—the long-legged Mary Ann. This original Edison bipolar generator, which had been slightly modified since its invention in mid-1879, had proven itself not only at Menlo Park but aboard the *Columbia*, where four generators continued to operate for over fifteen years.[20] The old stand-by thus provided the basic design for the most popular Edison generator, the "Z" dynamo. The Z, rated at 60 A (16-candlepower) or 120 B (8-candlepower) lamps, was installed in 71 of the first 99 isolated plants. In early 1882 the Isolated Company called upon Edison to produce larger models for more efficient operation in extensive installations, and the L (150 A lamps) and K (250 A lamps) dynamos began coming out of the

CATALOGUE AND PRICE LIST

—OF—

EDISON LIGHT FIXTURES,

MANUFACTURED BY

MESSRS. BERGMANN & CO.

292 to 298 AVENUE B, NEW YORK CITY.

These Electroliers, Brackets, etc., are especially designed for the Edison Incandescent Electric Lamp. They are provided with the standard sockets and wired in the best manner, in accordance with the requirements of the Board of Fire Underwriters and the rules laid down by the Engineering Department of the Edison Company. There is a large variety of designs of various prices, from which selections can be made suitable for all classes of work.

Most of the devices and fixtures illustrated in the following catalogue are manufactured and sold under patents which are controlled exclusively by the Edison Company and Messrs. Bergmann & Co., and the public are respectfully cautioned against all infringements of the same.

The illustrations in the catalogue represent only such leading styles of fixtures as its space permits us to show. It will be observed that the use of the Edison Incandescent Light offers a wider field for ornamentation in Electroliers, Brackets, etc., than that of gas. Special designs and estimates for all styles and classes of work will be furnished.

Goerck Street shops.[21] In addition, a smaller, 15-lamp dynamo, called the E, was installed in a few instances where only one or two rooms were to be lit. The increased capacity of the L and K dynamos came from the addition of one or two more pairs of field magnets. Their production did not require the extensive creative effort that went into the central station machines but did reflect the Machine Works' ability to adapt its products to changing needs.

The isolated systems spurred innovation in other ways. The fact that such systems, especially the smaller ones, were often operated by mechanics or engineers with little or no training in electrical equipment presented a challenge, for the new technology was liable to provoke considerable complaint if it did not perform as promised, even when failure was due not to faulty equipment but to improper handling. Edison responded by simplifying controls, meters, and indicators. He described one example in a letter to Batchelor dated December 31, 1881:

Bergmann and Company Catalog, 1883.
The rapid growth of the isolated plant business fostered the prosperity of S. Bergmann, a German-born mechanic who had worked for Edison in Newark before starting his own electrical manufacturing company in New York. Edison joined with Bergmann and Edward H. Johnson in forming Bergmann and Company to manufacture "special appliances connected with Electric Lighting," such as lamps, sockets, switches, and shades.

DISPLAY OF EDISON ELECTRIC LIGHT FIXTURES IN SHOW ROOM OF BERGMANN & CO.

Bergmann and Company Showroom.

This engraving from Electrical World *shows Bergmann's extensive display of Edison electric light fixtures. (Courtesy Smithsonian Institution)*

In putting out a great number of plants as we are now doing we have found it necessary to have a Regulator for the candle power of our lamps as the parties using the light are apt to run the lamps up very high and thus cause a great many breakages. Plus the average life would be shortened; great dissatisfaction caused people to get the impression that our statements as to life were not true. So I have devised an Indicator which works beautifully and I advise that hereafter all Isolated Plants shall be accompanied by one of them.[22]

The marketing of isolated plants also drove Edison and his colleagues to sharpen their commercial instincts and to adopt a more entrepreneurial attitude. Through the first few months of 1881, the Menlo Park demonstration system was kept aglow many evenings, and the Menlo Park staff were required to play host to both curiosity seekers and potential purchasers or franchisees. Soon after offices were established at 65 Fifth Avenue, Edison ordered some Menlo Park generators to be sent to New York, and the light was prominently displayed there. And subtler commercial touches were evident from time to time, as shown by Clarke's suggestion to Edison, dated November 11, 1881:

It would be well—as soon as the patterns can be spared—to round all the edges of the base and field for Z dynamo and introduce any features in the way of graceful curves which certainly will add much to the appearance and nothing to the cost of the machine.[23]

It is important to remember that, during the bold undertaking of the New York central station, the technical and commercial soundness of the Edison system was being demonstrated far and wide (in Europe and Latin America as well as the United States) by isolated plants, and that such plants for years remained the most important and common providers of the Edison light.

Edison's dreams, however, were wrapped up in the central station system, and it was to its creation that he primarily devoted himself upon moving to New York. The manufacturing plants in operation by the spring of 1881 made the isolated systems possible but had been brought into being to serve the central station effort. In the early spring the last legal obstacles were cleared away. On March 23 a contract was signed formalizing the relationship between the Edison Electric Light Company and the Illuminating Company. The contract gave the new company a license to construct and operate an electric lighting system based on Edison patents in two sections of New York City. The first was:

> . . . located in the lower part of the City of New York, bounded on the East by the East River, on the West by the middle line of Nassau Street, on the North by the middle line of Spruce Street, and on the South by the middle line of Wall Street.

The second section was simply defined as an "uptown" area to be designated in the future. Accounts differ as to how the First District was initially determined. William Hammer recalled many years later that he had, in the spring of 1880, gathered at Edison's request a number of large maps of New York City on which Edison shortly outlined the area for his first station.[24] By the time Edison moved to New York, certainly, the boundaries of the First District were set. During 1880, in fact, Edison had men surveying a number of streets in lower Manhattan to determine not only the amount of gas used for lighting in each building but also the power consumed in operating hoists (often mule-power) and other equipment. By the time the legal and financial details had been worked out for the Illuminating Company, Edison's knowledge of the First District and the kind of light and power market it represented was extensive.

The acquisition of this knowledge was consistent with the pains that Edison took to understand thoroughly the task facing him in New York. The surveys that began in late 1880 were logical extensions of the careful calculations with which Upton and others had occupied themselves earlier in the year. In December Edison wrote to the Light Company's

Map of Lower Manhattan Showing the Pearl Street District, 1882.

The First District, located in lower Manhattan, was strategically located near the city's financial and newspaper center. The almost completed Brooklyn Bridge is prominently featured on the map. Edison hired an Austrian electrical engineer, Hermann Claudius, for the purpose of "arranging, mapping and figuring out the main and subsidiary conductors" for the district. (Courtesy Smithsonian Institution)

Executive Committee to request the services of Herman Claudius for the purpose of "arranging, mapping and figuring out the main and subsidiary conductors for our first district in New York."[25] Using the information brought back by canvassers of the district and additional figures gleaned from gas company records, Claudius came up with block-by-block and house-by-house figures for the amount of gas potential customers consumed yearly, the number of lights they burned, and the amount of power they consumed for various purposes.[26] The information included in the district survey went into even more detail than this, for potential customers were asked about such things as the globes and shades they used on their lights, the use of gas for heating or engines, damage caused by gas impurities, complaints about leakages or excessive heat, various forms and uses of motive power, personnel required for tending engines, insurance rates, winter and summer hours of use, and so forth.[27] It is not clear how useful this detailed information proved to be in the long run, but it probably armed the Edison salesmen with some of their better pitches and may have guided Bergmann & Company in its development of auxiliary equipment. The essential facts of the First District were simply put—it contained about 1,500 gas customers, they used 20,000 gas jets, and, with the promise of free wiring and lighting costs in line with gas, most were ready and willing to receive the new light.[28]

On April 19, 1881, the New York City Board of Aldermen granted a franchise to the Edison Electric Illuminating Company to "lay tubes, wires, conductors and insulators, and to erect lamp-posts within the lines of the streets and avenues, parks and public places of the City of New York, for conveying and using electricity or electrical currents for purposes of illumination. . . ."[29] The two-page franchise resolution included appropriate clauses for insuring that the Illuminating Company repaired all damage to streets and pavements, that it did not allow its work to disturb other underground facilities and that it assumed full liability for all damage to private or public property. The good will that had been won from the aldermen during their visit to Menlo Park may have accounted for the fact that the city asked for only five cents per linear foot of trench as the fee for the street work. Mayor William Grace objected to the council's generosity but his veto was easily overridden, and Kruesi was able to begin laying his tubes by the end of April. The only other payment required by arrangements with the city were to the inspectors who were to ensure that the work and the restoration of the streets following it posed no public hazard. Edison remarked many years later that these gentlemen presented no obstacles, since they simply showed up each week to collect their fees and then promptly disappeared.

The laying of the mains in the First District took from the spring of 1881 through the summer of 1882, and, while it went forward without significant technical difficulties, the sheer amount of work involved in laying more than 80,000 feet of understreet conductors and the myriad of small problems needing rapid and careful attention made this the most exhausting of the tasks involved in completing the Edison system. In a matter of only a few months, Kruesi had the routine of the Tube Works and the street crews well established, and much of his energy was then devoted to seeing that the work met desired standards. He was given considerable responsibility for the day-to-day activities and the mechanical details of the underground system. Clarke, who was appointed the Illuminating Company's chief engineer, was responsible for many of the more technical aspects of the mains installation, particularly questions about electrical connections or conductor capacities. Edison himself, however, kept a close eye on things, and Clarke was careful to keep him informed and to involve him directly when changes were proposed, for example, reducing the number of fuses in house installations. Clarke wrote him:

You have verbally agreed to the abolition of a safety catch in the sockets and fixtures. Will you please state your agreement in writ-

Pearl Street District Statistics, 1880.

During 1880 Edison had men surveying the streets of lower Manhattan to determine not only the amount of gas used for lighting each building but also the power consumed in operating hoists and other equipment. Using this information and additional figures from gas company records, Claudius came up with block-by-block and building-by-building estimates of the number of lamps and amount of power potential customers would need.

Gaslight Use.

ing so that should the matter come up later I can shew that you have been consulted, although the change is to be made on my authority.[30]

A little later Clarke wrote in more general terms:

It is well to have all steps involving a departure from the old system properly stated in writing, so that the responsibility can be

Have you any machinery driven by foot power? 43

Nonlighting Power Consumption.

Yes	No
1 for spinning wheel.	⊹⊹⊹⊹⊹⊹⊹⊹⊹⊹⊹⊹
1 " Bellows for flowers.	⊹⊹⊹⊹⊹⊹⊹⊹⊹⊹⊹
1 " Cutter	⊹⊹⊹⊹⊹⊹⊹⊹⊹
1 " "	95 No
1 " 30 Lathes.	
1 " 4 machines	1 – 6 lathe.
1 " 2 "	1 – 1 press
1 " 4 stovepipe	1 – 4 "
machines	1 – sewing machines.
1 " stamp machine	1 – 6 " lathes.
1 " 5 lathes	1 – Embossing machine.
1 " 1 "	1 – 4 lathes.
1 " 2 "	1 – 2 lathes.
1 – 4 press	1 – for Bellows
1 – 1 "	1 – sewing machines,
1 – 1 "	1 – lathe
1 – 1 Copper cutter .	
1 – presses	33 parties Yes.
1 – 2 lathes .	
1 .	
1 . spindle	
1 . 6 machines	
1 – 2 lathes	

placed, and it can be ascertained if due discretion has been used and proper parties consulted.

I wish you however to bear witness if you can to the fact that I am painstaking with reference to the system generally and in detail, that I take particular attention to consult you on all points pertaining to your system and never take or have taken the initiative without your assent. Can you do this? [31]

Numbers of Gaslights.

Block	Street	House	Number of Burners	Block	Street	House	Number of Burners
	Number of Burners from 8 A.M. to 5 P.M. in Winter.						**221**
10	Spruce street	2,4,6	2		Corner Gold & Beekman st.		305
	"	8	150				6
	"	12/14	8				6
			160		Beekman street	56	4
	Williams st.	179	6		"	54	4
	"	177	4		"	46	1
	"	175	2		"	44	10
	"	171	1		Corner Beekm. & Williams street		4
	Cor. William & Beekman street		2				23
			15		Williams street	176	2
	Beekman street	36	2		"	178	2
	"	34	2				4
	"	32	8				
	"	30	22	25	C. Ferry and Gold street		3
	"	28	8		Ferry street	10	1
	"	24	23		"	26/28	1
	"	20	3		C. Ferry & Clifford street		2
	"	14	1		Clifford street	67	2
	Corner Nassau & Beekman street		2		"	65	9
	Nassau street	140	1/2				11
		142	5		Beekman street	88/90	12
		144	1		"	84	3
		146	1		"	82/84	14
		148 / 146–148	2		"	76	8
		150–152	25		"	74	10
		152	8		"	72	6
			44		"	70	8
					C. Gold & Beekman street		1
20	Spruce street	26	1				62
	"	28–30	4	28	Gold street	50	1
	"	34	4				1
	"	36	4		Cor. Ferry & Clifford street		1
	Corner Spruce street & Gold street		2				1
			15		Pearl street	301	5
			305				420

While men with more of an engineering mentality, like Clarke, Kruesi, or central station engineer John W. Lieb, assumed a greater role in the development of the Edison system as it grew more and more complex, it was still Edison who had the ultimate responsibility for the new technology.

At the same time that work was starting on the street mains in the First District, the locating of a site for the generating station assumed a major priority. In later recollections, Edison spoke of his frustration over high Manhattan real estate prices, even in an area selected because it accommodated industry. In truth, once the boundaries of the district had

Laying the Electrical Tubes

been set, Edison did not have a wide range of options in locating his generators. The planned distribution system had a range limited to about a half-mile from generators to the farthest lights. In a system depending on street mains, this required the generating station to be as close as possible to the center of the system area. The choice of the buildings at 255 and 257 Pearl Street, therefore, was probably determined not simply by relative cheapness (as Edison claimed) but also by geographical requirements. In early May 1881, the Illuminating Company completed the purchase of the Pearl Street buildings for $65,000.

The property had originally been used for commercial activity, so the building at 257 Pearl Street, that would house the heavy generators, steam engines, and boilers, had to be substantially strengthened before machinery could be installed. There was initially some concern about the

**Laying Electrical Tubes,
June 24, 1882.**

This Harper's Weekly *engraving shows the enormous task involved in laying mains in the First District. From the spring of 1881 to the summer of 1882, workers laid more than 80,000 feet of conductors under the streets.*

ability of the second floor to carry the load of the huge dynamos, but Clarke had the interior reinforced with the best wrought iron, and, only a few weeks before the station began operating, had the load-carrying capability certified by engineers. The building was 25 feet wide and 100 feet deep, with four stories and a basement—far from Edison's ideal for his model station but, as it turned out, perfectly adaptable for his needs. The basement was used for storage of coal and the removal of ashes. A conveyor was installed to carry coal up to the first (ground) floor, where four Babcock & Wilcox boilers provided steam at 120 pounds per square inch pressure to the generators on the next floor up. On the floor above the generators was a large bank of lamps used to test dynamos and measure the station's load. The building at 255 Pearl Street was kept for equipment storage, offices, sleeping quarters for station attendants, and testing and measuring facilities. The station of the Edison Electric Illuminating Company did little to improve the ambience of its dilapidated neighborhood but proved to be a very functional structure in its almost twelve years of service.

In the spring of 1881 Edison was still driven by the challenge of developing a generator suitable for the kind of central station he envisaged. The winter's test of the large Menlo Park 100-horsepower machine was considered a success, but design modifications were necessary. The first task of the Machine Works on Goerck Street, in Edison's mind, was to build the new large dynamos that would be the mainstay of central stations. The Menlo Park machine had been directly driven by a Porter-Allen steam engine, but Edison was not fully satisfied with its performance, especially its speed regulation. He thus asked the Armington & Sims company to provide, for $2,000, a 125-horsepower engine capable of running at 350 revolutions per minute.[32] Meanwhile, tests continued to be made of the suitability of other engines, and Charles Porter was also contacted for possible further work. Edison considered the behavior of the engine linked to his dynamo to be a critical factor and continued to be less than fully satisfied with the machines he could get. He kept his options open even as he found it necessary to commit himself to the large dynamos nearing completion. This proved fortunate, for supply difficulties kept Gardiner Sims from doing all the work Edison needed for Pearl Street. However, when the Porter-Allen engines were put into operation at the station, the deficiencies of the speed-regulating governors made it impossible to run more than one engine at a time without setting up horrendous vibrations in the dynamo room. The Pearl Street station operated on only one generator for four weeks until Edison devised a mechanical linkage between the steam engines.[33] Shortly afterwards the Armington & Sims engines were installed and multiple running of the engines was much easier. Direct-connected steam engine-dynamos did not, in fact, turn out to be the most satisfactory form of power plant for the early central stations, and most stations after Pearl

Street returned to the use of smaller belt-driven dynamos such as had been used at Menlo Park.

The design and construction of the large dynamo's steam engine, while a matter of great concern to Edison, was not something he could manage directly. The building of the dynamo itself, however, was in his hands, and much of his creative efforts in 1881 were bent toward improving the large dynamo tested at Menlo Park early in the year. Even at this point Edison experimented with some radical design changes, such as a disk armature that could be operated at very high speeds (over 1,200 revolutions per minute) and would eliminate the iron core used in the drum-wound armatures of other machines. The copper-disk dynamo he had built at Menlo Park early in 1881 worked and was patented (U.S. Patent 263,150) but was not a practical point of departure for the central station machine.

Soon after the Goerck Street shops were set up, work began on a second large dynamo intended for use in the Edison exhibits at the upcoming Paris International Electrical Exposition. This machine was soon designated the "C" model and was the first of twenty-three such machines built at Goerck Street. The Paris machine (as it was called at the time) differed from the experimental Menlo Park dynamo primarily in the construction of its armature. One of the sought-after advantages of the large dynamos, not satisfactorily achieved with the Menlo Park machine, was the lowering of the resistance in the armature due to its size. The C dynamo was built after months of experimentation on armature construction, Francis Jehl being called up from Menlo Park in May to set up a testing room at Goerck Street to measure armature performance. All summer the Machine Works mechanics struggled to build a machine that minimized armature resistance and yet did not burn out or spark while running at speeds of over 300 revolutions per minute. Working on such a large and complex machine presented new and difficult problems and required considerable toil. Jehl noted once, for example, that it took fifty-five men working eight days and nights to change the construction of the armature when one of its parts had to be reshaped.[34] By the end of the summer, however, the armature resistance had been lowered to less than one-hundredth of an ohm and most of the problems of cooling the machine (using an air blast) and regulating its operation had been solved. In September the giant dynamo was disassembled, packed into 137 crates, and loaded aboard the ship that had recently brought P. T. Barnums's famous elephant Jumbo to America, giving the machine the name by which it was known familiarly ever after. The cost of the work must have dismayed even Edison (who could be quite cavalier about such things), for he lost little time in writing to the Light Company seeking compensation—$6,171.31 for experimental costs alone.[35]

The machine that emerged from this labor was a technical triumph. It far surpassed in size any electrical machine ever attempted before. The

**Edison's Disk Dynamo,
May 20, 1881.**

*Edison's drawing, made in prepara-
tion of a patent application, shows
his disk dynamo, which evolved from
attempts to improve the large dynamo. The disk armature, a radical
design change, could operate at very
high speeds (1,200 or more revolutions per minute) and eliminated the
iron core used in the drum-wound
armatures of other machines. Although Edison obtained a patent, the
disk dynamo was not practical for
central stations.*

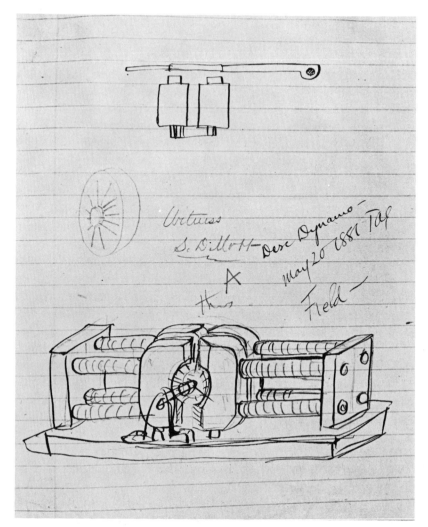

Jumbo measured 168 by 105 inches at the base and weighed just over 30
tons. Armington & Sims made the steam engine driver for the machine
that went to Paris. When operated with the air blast at 350 revolutions
per minute, the dynamo produced a current of about 500 amperes at 103
volts (a power of 51,500 watts) and could light as many as 700 A-lamps.
at the Paris Exposition a smaller load, about 500 lamps, was used and
the air blast was not required. The reception at Paris was everything
that could have been hoped for; Batchelor cabled home that Edison's generator and light had carried away the highest honors, whereas rivals
such as Hiram Maxim and Joseph Swan had received lesser prizes. The
Exposition success elevated the standing of the Edison system in the
eyes of European scientists, engineers, and businessmen, and the Paris
awards carried much prestige on the other side of the Atlantic as well. A
considerable boost was given to the commercial enterprises on the con-

EDISON'S STEAM DYNAMO-ELECTRIC MACHINE AT THE PARIS ELECTRICAL EXHIBITION.

tinent that Batchelor had been chosen to organize, and, when the Exposition was over, the dynamo was installed in the factory at Ivry-sur-Seine of the Société Continentale Edison. Contracts for isolated plants in showcase locations, such as the Opera Houses in Paris, Berlin, and Milan, were quickly arranged, and the Edison system in Europe was off to a spectacular start.

The European success continued into 1882, for while Batchelor worked in France to get manufacturing under way, Edward Johnson and William Hammer went on to London for an electrical exhibition at the Crystal Palace. There they received the second Jumbo dynamo from Goerck Street, a bit larger than the Paris machine, with an Armington & Sims 200-horsepower engine. This large dynamo was not intended for the Crystal Palace display, which was powered by twelve smaller Z dynamos, but for installation in a model central station—far less ambitious than the Pearl Street station, to be sure, but no less strategically situated for gaining the attention of influential individuals and institutions. Since the legal obstacles to installing a system of underground mains in London were even more formidable than in New York, and since it was

Jumbo Dynamo at the Paris Electrical Exhibition, 1881.
Edison's electric light and generator (using an Armington and Sims steam engine) carried away the highest honors at the Paris Electrical Exhibition. This success elevated his lighting system's standing in the eyes of European scientists, engineers, and businessmen and gave a considerable boost to the commercial enterprises Batchelor organized in Paris for marketing the light in Europe. (Courtesy Smithsonian Institution)

Edison Exhibit at the London Crystal Palace Exhibition, 1882.

The Edison display at the Crystal Palace was a small system powered by twelve Z dynamos. A larger demonstration system installed on the Holborn Viaduct used two Jumbo dynamos. (Courtesy Smithsonian Institution)

desired to install the London system quickly and at minimal cost, the station was located on the Holborn Viaduct, which crosses over Farringdon Road at the western boundary of London's "City," the financial and communications center of the British Empire. The Viaduct allowed electrical conduits to be installed underneath, obviating underground excavation and the need for permits under the gas statutes.

On January 12, 1882, the generator at 57 Holborn Viaduct began operating. In April it was joined by a second C dynamo and the two operated together in much the fashion envisioned for the Pearl Street station. The demonstration was an immediate success. The combination of the impressive display that Hammer designed for the Crystal Palace and the highly visible demonstration of street and domestic lighting by the Holborn Viaduct station had a considerable impact. The London

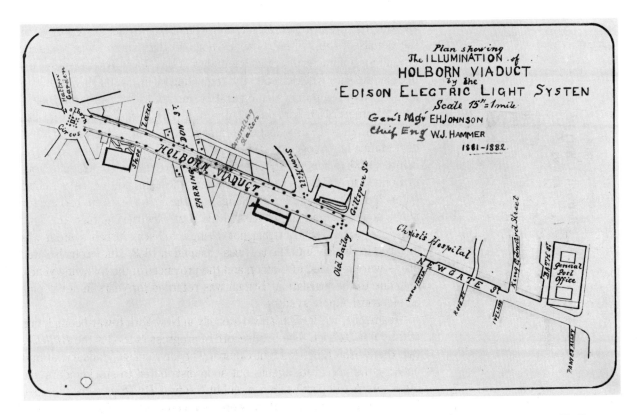

press reacted favorably, despite initial skepticism. Reporting on the Crystal Palace display, the *Daily News* wrote:

> There are two questions to solve besides the production of a lamp, viz.: the proper distribution of electricity through a town, and its economy relative to gas. Mr. Edison is far and away in advance of all rivals in the solution of these problems. His exhibition is the wonder of the show, and his representative is certainly the prince of all showmen. There is but one Edison and Johnson is his prophet. One feels after an hour with Mr. Johnson that there is nothing left to be done, that one's gas shares must be sold at once, that there is only one system, and that is Edison's and that every question has been solved.

The Holborn Viaduct system was indeed a full-scale demonstration of the Edison light. The installation stretched a half-mile along the Viaduct, from Holborn Circus to the General Post Office. By April a total of 938 lamps were wired up, including 164 street lamps of 32 candlepower, a new product of the lamp factory that allowed Edison to demonstrate further the versatility of his system. Customers included hotels, restaurants, shops, offices, and the City Temple, in addition to a portion of the General Post Office itself.[36] The newspapers of Fleet Street were nearby,

Plan for Lighting the Holborn Viaduct, 1881–1882.
The Holborn Viaduct project was intended as a temporary demonstration, not a permanent commercial station. By choosing the Viaduct, Edison's London agents were able to install the station quickly and with minimal cost because the electrical conduits could be hung underneath without excavations or permits under the gas statutes. The Viaduct was a testing ground for several key elements of the Pearl Street system, especially safety devices and regulating mechanisms. (Courtesy Smithsonian Institution)

allowing the press a favorable vantage point from which to report on all the details of the system. The Holborn Viaduct station was a testing ground for a number of critical elements of the Pearl Street system, especially the safety devices and the regulating mechanisms. It was always intended as a temporary demonstration and not as the basis of a permanent commercial station. Installation of a permanent, Pearl Street-like system in London was in fact contemplated even before the Holborn Viaduct station began operating. Edison wrote to Johnson at the end of December 1881 to suggest that he "keep your eye out on some square mile of London in which there is a slum near the center in which we could obtain a building cheap."[37] Not long after the Edison system had proven itself, however, Parliament passed the first Electric Lighting Act, with franchise terms so strict that not a single full-scale urban system was installed in Britain until the law was changed in 1888. Numerous isolated plants were installed in London and the provinces in the following years, but, due to the legislation, Britain was retarded for years in developing an electrical supply system.

Meanwhile, work went ahead steadily in New York toward completion of the First District. The best record of progress in 1882 is the *Bulletins* that the Edison Electric Light Company began issuing in January, at first only for the use of its agents but soon distributed to stockholders and other interested parties. Since all but the first *Bulletins* were intended for public consumption, they were not completely reliable regarding problems encountered in the last stages of the work, but the First District activity really allowed little room for dissembling in any case. The first *Bulletin*, dated January 26, reported on "Lighting up New York City" as follows:

Between six and seven miles of street mains have thus far been laid in the down town district. The bad weather has caused a suspension of laying mains for nearly a month. About six miles more mains must be laid. The third mammoth dynamo has been completed. . . . Mr. Edison is satisfied with the improvements in these dynamos and will now hasten the work on the uncompleted dynamos for the First District in this city. Six will be finished first, and after they are started in the Pearl Street building, another six will be finished to be placed in the adjoining building, which also belongs to our company. The meter to be used in the first district is completed and satisfactory. It registers with almost absolute exactness. This gives still another advantage over the existing gas system, where the meter question is one of looseness and uncertainty. No time for lighting up the Down Town District can be fixed. The work is being pushed forward with the utmost vigor, but the undertaking is so great, probably a few months must yet elapse before the district is actually lighted.

The meter had, in fact, posed difficult problems, and it is not clear that these had been successfully dealt with by January 1882. By the late spring, the months of experiment had produced a chemical meter using pure zinc plates in a solution of zinc sulphate. A shunt of German silver (a copper-zinc-nickel alloy) with a precisely adjusted resistance diverted a fraction of the current from the circuit into the meter where it caused zinc from the solution to deposit on the plates. The total current consumed by the customer was determined by periodic collection and weighing of the zinc plates. Where meters were exposed to cold, a light bulb was installed in the meter box with a thermostatic switch to turn it on to keep the solution from freezing in frigid weather. Edison was quite proud of his meter and defensive about its accuracy. The meters for Pearl Street were manufactured by Bergmann & Company and installed in houses as they were wired into the system. They were not, however, used as the basis for billing until the Pearl Street station had been operating for almost six months, since, as the Electric Light Company's 1883 *Annual Report* stated, "Mr. Edison was continually making changes and improvements . . . and the Illuminating Company wished to avoid being tied up by contracts to furnish light, until after Mr. Edison had entirely completed his observation."

In March the fourth and fifth *Bulletins* reported that, with the spring thaw, Kruesi had resumed laying the underground conductors and was moving ahead at a rate of a thousand feet a day. By mid-April it could be reported that the building at 257 Pearl Street had been fitted with boilers and auxiliary equipment, six Porter-Allen engines had been received, and work on the dynamos was nearing completion. The next couple of months brought word of steady progress on the largest outstanding task, completion of the street mains—12,500 feet were laid in April, another 7,923 feet in May. The twelfth *Bulletin*, dated July 27, reported "the entire network of underground conductors finished, aggregating over 80,000 feet." Further work consisted largely of completing connections between mains and houses and installing meters and fixtures. Of special significance was the report that the New York Board of Fire Underwriters had given its approval to the Edison installations and indicated that their presence would have no adverse affect on insurance rates, provided they were properly insulated. Finally, the completion of the work at the station itself was noted:

> The equipment of the central station in Pearl Street is also finished. Fire was built under the boilers for the first time on June 29th, and on the next day the small engine used for the coal conveyors, blowers, etc., was started and all that portion of the equipment was found to work well. The first steam dynamo was started July 5th; and, July 8th, a satisfactory experiment was made on 1,000 lamps arranged on an upper floor. Since that date, some of

Edison Chemical Meter, 1882.
By late spring, months of experimen-
tation produced a chemical current
meter using pure zinc plates in a so-
lution of zinc sulphate. A wire shunt
made of German silver (copper-zinc-
nickel alloy), with a precisely ad-
justed resistance, led a known frac-
tion of the current from the circuit
to the meter, where it caused zinc
from the solution to deposit on the
plates. The current consumed by
the customer was calculated from
the amount of zinc deposited, deter-
mined by periodic collection and
weighing of the plates. (Courtesy
Smithsonian Institution)

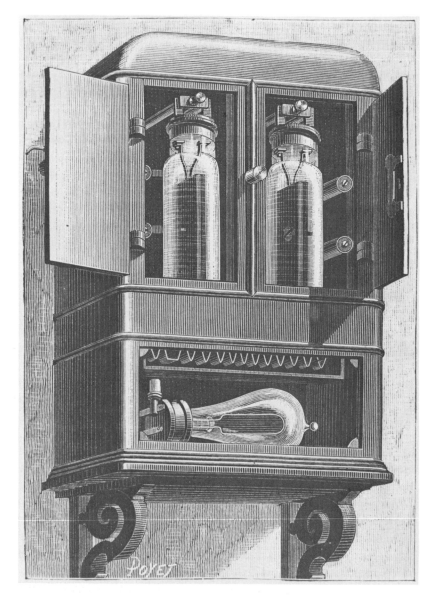

the other engines and dynamos have been carefully tested with the 1,000 lamps, and the details of their adjustment perfected. The field regulating apparatus has also been tested, and the electrical indicator, the first ever used on so large a scale, has also been found satisfactory.

Only a few final tests of the electrical system remained. A month later, in the thirteenth *Bulletin,* dated August 28, all was said to be ready, and, indeed, "a number of buildings in various parts of the district have already been lighted." Connections to houses continued, 226 having been

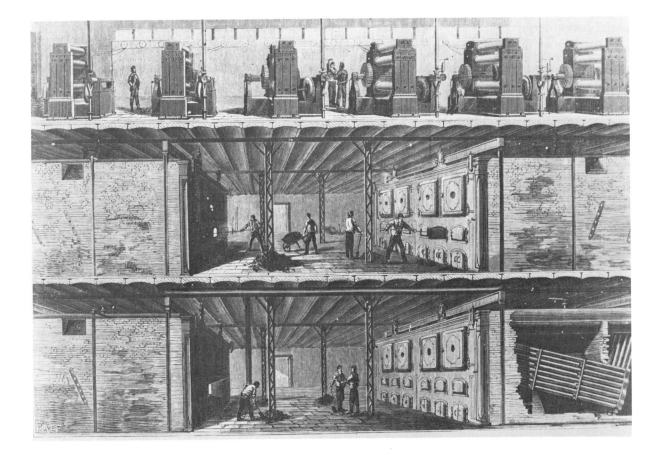

put on the system as of August 26. "No serious obstacles" were anticipated before "the early lighting up of the entire district."

Every effort was made to test the installation before any publicly announced start-up. The system was too complex and too much was at
stake to allow the debut of Pearl Street to be left to chance. Feeders and
mains were tested with one of the Jumbos connected to a small portion
of the system at a time. There were, station engineer John Lieb pointed
out later, no proper meters in the entire station, but indications of a
rough sort were devised to show when loads were high or low, and resistances were plugged in and out of the feeder circuits to control voltage
levels. Most of the last-minute worry revolved around the integrity of
the underground conductors. No serious problem was encountered, but
one incident attracted brief press notice, somewhat to Edison's dismay.
On a day in late August, when the Jumbo was being tested and a considerable portion of the mains were in the circuit, word came of a leakage of
current near the corner of Nassau and Ann Streets. There a crowd
gathered around a wet spot in the road, for whenever a horse passed
over it, the current from Pearl Street gave the animal a surprising shock,
startling cart drivers. It was concluded that a worker must have driven a

**Interior of the Pearl Street
Station, 1882.**

*By the end of June the installation of
equipment in the Pearl Street station
was completed and the boilers were
fired up. The first dynamo was started
on July 5, and other equipment was
tested during the rest of the month.*

spike through a main tube and the soaked ground conducted the current to the surface. This resulted in little more than mildly embarrassing publicity but kept Edison and the Illuminating Company on their guard.

By September enough testing had been completed to give Edison the confidence to begin service at Pearl Street, although all connections in the system had not yet been finished. On September 4 Edison synchronized his watch with John Lieb's and, accompanied by Kruesi, Bergmann, and others, made his way to the offices of J. Pierpont Morgan in the Drexel building at Broad and Wall streets, at one edge of the First District. There, he supervised the installation of the safety catches and at three o'clock in the afternoon turned on the office lamps. It was an understandably dramatic moment which, as so often happened in the story of the electric light, was caught most eloquently by the press. The best description of the general impression made by the lights in the portion of the District illuminated on September 4 (bounded by Nassau, Wall, Pearl, and Spruce Streets) is from the *Herald:*

> In stores and business places throughout the lower quarters of the city there was a strange glow last night. The dim flicker of gas, often subdued and debilitated by grim and uncleanly globes, was supplanted by a steady glare, bright and mellow, which illuminated interiors and shone through windows fixed and unwavering. From the outer darkness these points of light looked like drops of flame suspended from jets and ready to fall at every moment. Many scurrying by in preoccupation of the moment failed to see them, but the attention of those who chanced to glance that way was at once arrested. It was the glowing incandescent lamps of Edison, used last evening for the first time in the practical illumination of the first of the districts into which the city had been divided. The lighting, which this time was less an experiment than the regular inauguration of the work, was eminently satisfactory. Albeit there had been doubters at home and abroad who showed a disposition to scoff at the work of the Wizard of Menlo Park and insinuate that the practical application of his invention would fall short of what was expected of it, the test was fairly stood and the luminous horseshoes did their work well. . . .

To a *Sun* reporter, Edison remarked simply, "I have accomplished all I promised."

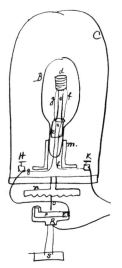

CHAPTER 8

Afterword

Such minds resemble a liquid on the point of crystallization. Stirred by a hint, crystals of constructive thought immediately shoot through them. That Mr. Edison possesses this intuitive power in no common measure is proved by what he has already accomplished. He has the penetration to seize the relationship of facts and principles, and the art to reduce them to novel and concrete combinations.
—John Tyndall, January, 1879

When an abnormal man can find such abnormal ways and means to make his name known all over the world with such rocket-like swiftness, and accumulate such wealth with such little real knowledge, a man that cannot solve a simple equation, I say, such a man is a genius—or let us use the more popular word—a wizard. So was Barnum! Edison is and always was a shrewd, witty business man without a soul, an electrical and mechanical jobber, who well understood how to "whoop things up," whose only ambition was to make money and pose as a sort of fetich for great masses of people that possess only a popular notion of an art, and who are always ready to yap in astonishment at some fire-work display that is blown off for the benefit of mankind.
—Francis Jehl, ca. 1913

Thomas Edison's "method" has been the subject of wonder, comment, and analysis at least since the popular press identified the man as a phenomenon in the late 1870s. The first book-length biography of Edison was a popular work by James McClure, issued early in 1879, and the literature that has accumulated since is as impressive in size and range as Edison's final list of 1,093 patents. The comments of Tyndall and Jehl are good representatives of the spectrum of opinion about the nature of Edison's genius. The sober British physicist was well equipped by temperament and social setting to be judicious in his views of the surprising American; his comments convey something of the wonder with which many of Edison's contemporaries viewed him. The caustic views of Jehl, the former helper from Menlo Park, take on a special poignancy in light of his later role as the chief custodian of the Edison legend in the shrine erected to it by Henry Ford at Dearborn, Michigan. The vituperation of the remarks he expressed in middle age is hard to explain, but others for whom the streak of humbug in Edison's character overwhelmed everything else have echoed similar sentiments.

It would be surprising indeed if a man who dominated his sphere of life as thoroughly as Edison did not arouse controversy over the nature of his achievement. In Edison's own day, much of the controversy was simply commercial rivalry as competing inventors and businessmen pursued their own interests in claiming prior invention or illicit infringement. Courtroom patent battles, heated exchanges in both the popular and technical press, and even bitter personal conflicts were the expected lot of an active inventor in late nineteenth-century America, and Edison had more than his share. The electric light, a complex technical goal pursued by many men for several decades, entangled Edison and his colleagues in numerous legal and commercial fights with a variety of competitors in the decade after 1879. These fed some lively arguments about who had accomplished what, both at Menlo Park and elsewhere.

Some of these very public arguments involved people who had worked closely with Edison. Frank Pope, who helped Edison get established in his first years in New York, later devoted much effort to denigrating Edison's work on the electric light. A few of the old gang from Menlo Park voiced the same disaffection shown by Jehl in later years, the most prominent example being glassblower Ludwig Boehm. Several individuals, such as Frank Sprague, Nikola Tesla, and Edward Acheson, who later became famous for their own technical contributions, spent time in the Edison shop, only to complain later of an atmosphere that stifled their own creative bent. It was also claimed that Edison freely took of the ideas and talents of others without regard for proper credit or reward.

Such opinions never tarnished the Edison myth, despite continuing efforts of rivals and uncomplimentary writers. The valuable assistance of key individuals at Edison's side, at Menlo Park and later at his West Orange laboratory, was always readily acknowledged, both by Edison himself and by his biographers, adulatory and otherwise. Edison retained the image of the indispensable man in all that went on in his laboratories and in the creation of the inventions and systems still linked to his name.

This image survives also in scholarship that continues to define and interpret Edison's life and work. To be sure, attempts have been made to give a different coloration to the enduring image. For example, the most recent Edison biography, Robert Conot's *A Streak of Luck,* emphasized some of the less attractive sides of Edison's personality—his vulgarity, boastfulness, and sloppiness. Conot, however, also brought into question the true source of Edison's carbon lamp. Citing a Menlo Park notebook reference to a report of Joseph Swan's low-resistance carbon lamp in England, Conot argued that the report was the source of Edison's renewed interest in carbon in the early autumn of 1879. The weakness of this argument is spelled out below in the Bibliographical Note, and there seems little reason to consider seriously any outside origin for the carbon filament idea. Nevertheless, Conot's conclusions would not sound out of place in any earlier biographies: "Between 1878 and 1882 Edison constructed the prototype for the entire electric light and power industry. . . . Edison threw open the door through which not only he but a host of scientists and inventors rushed to make discovery after discovery. . . . Edison was a scientific explorer who discovered new continents where prevailing opinion held that none existed."[1] Whatever blemishes may have been highlighted in the picture of the folksy old inventor, the aura of genius is hard to dispel.

The more important scholarly interpretations of Edison's work come not from his biographers but from the historians of technology who have attempted to incorporate the past century's premier inventor into their models of technological change. Abbot Payson Usher, for example, in the 1954 revision of his classic, *A History of Mechanical Invention,* included Edison's electric light among the numerous instances of "acts of insight" that he cited in his analysis of the emergence of technical novelty.[2] To Usher, three elements were crucial to the successful invention of the incandescent light: (1) a high lamp resistance, (2) solving the problem of occluded gases, and (3) the use of carbon for the filament material. This is a reasonable analysis, but Usher is less successful in explaining the significance of these elements. He declares, for example, that "high resistance was essential to adequate illumination," which is not only untrue but misses the true importance of high resistance for parallel circuits. The primary act of insight in the invention of the lamp was, to Usher, the realization in the late summer of 1879 of the possibility of adapting carbon as a filament material. All the previous work is seen as "setting the stage" for the final solution of the lamp problem, and all subsequent work as the solution of subsidiary technical problems and necessary further refinement and development. While apparently aware of the complexity of Edison's invention, Usher does not fully appreciate the extent to which the inventive activity itself reflected that complexity.

Neither the work of Usher and economic historians who followed in his footsteps, nor that of most other historians, such as Harold Passer and Arthur A. Bright,[3] who focused more narrowly on electrical technology, has gone much beyond the traditional stories of the light's invention and subsequent development. This is not to denigrate their contributions, for they have helped to integrate the elements of technological change into a broader historical context. For the most part, however, they have relied on the sketchy and unreliable picture of Edison's work that emerges from reminiscences and contemporary reports. Furthermore, their generally limited concern for the actual processes of invention has led to superficial and uncritical references to inventive activity.

In the last decade or so, a new generation of scholars with greater interest in the detailed processes underlying technological change has contributed analyses that go beyond those of older historians. Articles by Christopher Derganc, David Hounshell, and George Wise, for example, reflect this more sophisticated inquiry into Edison's methods and style and how they related to the social and professional setting around him.[4] Of greatest influence, however, is the work of Thomas Hughes, whose studies of Edison's work and the subsequent development of electric power technology culminated in his *Networks of Power.*[5] Through all his treatments of Edison, Hughes has emphasized Edison's concern with systems rather than mere technical components. This perception of Edison's work is, as we have seen, far from novel. From the press's first announcement of Edison's interest in the electric light to the most recent biographies, few serious observers have failed to point out the systemic nature of the electric light and power technology that Edison set out to create in 1878. Nor has Edison's main goal been entirely missed by those who have described his approach. If Hughes goes beyond earlier observers, it is in ascribing an overall "systems approach" to Edison's inventive activity.

It is not obvious how a systems approach differs from simply inventing systems or inventing within the context of systems. Picking one's way through Hughes's peculiar terminology, one surmises that working with systems enhanced Edison's ability to spot the key weak points in a technology. The solution of problems at those weak points, according to Hughes, would lead to a general improvement of the system and to the disclosure of new strategic weak points, which would, in turn, be the object of the next intensive inventive effort. This process, it seems, would continue until the technological system was ready for implementation (and often further as the system evolved). Edison, it is claimed, was attracted to this approach because working with systems allowed him to control every aspect of the development and application of his inventions.[6]

As we have seen, Edison's comprehension of the systems needs of the electric light was not immediate. At the start, in the fall of 1878, he was excited by the possibility of solving the problem of regulating the action of a platinum lamp. Such a lamp was seen as fitting into circuits powered by generators like those used by William Wallace. In his comments to the *New York Sun's* reporter in mid-September, Edison described how he could light up all of lower Manhattan with a 500-horsepower engine and Wallace's generators, thanks primarily to the lamps he expected to have ready in a matter of weeks. To be sure, Edison's vision, even at that point, incorporated the widespread distribution of electric power from a centralized source, the use of underground conduits and adapted gas fixtures, and parallel circuits to allow independent control of individual lamps. The full technical requirements of this system, however, were realized only after long months of experiment and trial. After five or six weeks, for example, Edison began putting some of his assistants to work on developing an appropriate generator, but even this decision owed more to dissatisfaction with the generally crude state of generator technology in the late 1870s than to the rejection of specific available generators for the particular system Edison had in mind. Whatever attraction systems may have had for Edison, at least at this stage of his career as an inventor, was born less from a "systems approach" than from an enormous ambition and a supreme confidence in his own abilities.

The extent to which Edison was able to solve the myriad technical problems involved in making his system workable owed more to the tremendous advantages he possessed in his Menlo Park establishment and the capable men he gathered around him than to the completeness of his vision. The other men, like Moses Farmer, William Sawyer, and Joseph Swan, who tried to make a practical incandescent lamp, had little opportunity to demonstrate a comprehension of the systems requirements of their own inventions because their lamps never worked well enough. Edison, however, attacked these requirements relatively early, not necessarily because he started with a clearer and more comprehensive concept of an electric light and power system, but because he was stimulated by the confident expectation of the public and the business community, as well as his own, that a workable and complete technology would emerge from Menlo Park, and was backed by a great deal of money. This combination of expectations, resources, and confidence led Edison to leap ahead in designing those elements of an electric lighting system that other inventors never even had the chance to deal with. Edison was the first to devise successfully the single element that all agreed was the key to incandescent electric lighting, the lamp, which then allowed the full elaboration and installation of his system. The completeness of that system was more the product of opportunities afforded by technical accomplishment and financial resources than the outcome of a purposeful systems approach.

We ended our story with the opening of the Pearl Street Station. This was, of course, not the end of Edison's work on electric light and power. For the next ten years, the extension and perfection of his system occupied portions of his time but in a very different setting from that of his first four years of pioneering. In the New York City shops and offices, under the auspices of the Edison Electric Light Company, and then, after 1886, in the fine laboratory complex he built in a valley in West Orange, New Jersey, Edison continued to pursue solutions to technical problems presented by "his" system. It was a great tribute to his initial success, however, that the electric light and power system that began its rapid spread throughout America and Europe in the 1880s did not long remain his. Not only did a host of outside rivals and competitors quickly emerge, many of them, like Charles Brush and Elihu Thomson, technically astute and creative, but the number and complexity of problems to be solved in the expanding Edison enterprises soon exceeded the ability of even an Edison to comprehend and manage. The solution of these problems, by Edison, his coworkers, and his rivals, is itself a story worth telling, and, indeed, scholars have begun to piece together its outlines and some of its details.[7] It is the story of the creation of our modern technological order based on central electric power generation and the transmission and distribution of that power to every corner of life.

Edison's work in Menlo Park and New York City between 1878 and 1882 was the central element in the origins of this great technological transformation. But this it not our subject here. Our aims have been more modest: to describe and comprehend the work itself, to understand the patterns and features of the inventive act. The incandescent electric light, for all the complexity of the system it bred and the implications it held for the future, was above all a product of creativity, of the ingenious application of men, resources, and ideas to the forging of something new. In becoming too absorbed with analyzing and understanding technological change, in formulating and applying a useful language for describing such change, we risk forgetting the human roots of innovation—the urge to create something that has never existed before, something that will be admired for its ingenuity, appreciated for its usefulness, and valued for its contribution to human well-being. These roots run deep throughout historical experience but flourished and gave fruit as never before in the nineteenth century. The social and intellectual conditions that made this possible were the products of a whole host of circumstances, ranging from the rise of modern science and an accompanying materialism to the great opportunities of frontier and empire that shaped American and European destinies. An invention can be fully understood only as an artifact of individuals and their times.

Partly for this reason, most writing about significant inventions has been in the framework of biography. We have chosen a different approach here, for, as important as the context of an individual's life is for viewing an invention, that context is only part of what there is to see. This is especially true with the incandescent electric light although it is partially obscured by the larger-than-life figure of Edison, who stands astride the light's creation as though it were all merely an extension of himself. It detracts nothing from Edison to declare that his genius did not work alone or without shortcomings and setbacks. To say only that, however, as this study makes clear, is just the beginning of understanding the invention itself. That understanding starts with defining the place of the individual in the inventive process and proceeds to close scrutiny of the process itself.

To the task of inventing the electric light Edison brought the greatest electromechanical talent of his generation. Lord Kelvin's explanation why Edison's feeder-and-main system had not been thought of before, because "no one else is Edison," epitomized the contemporary feeling that Edison brought something unique to his work, not possessed by anyone else. Edison's fame for technical adroitness was already great by 1878, resting on his solution of truly difficult technical problems, as in multiple telegraphy, and on the creation of simple but ingenious devices, such as the carbon telephone transmitter and the phonograph. The records of work on the electric light show time and again the sureness with which Edison moved in the realms of the electric circuit or the electromechanical relay. The electric light posed very new problems in these and other areas, but, proceeding by analogy and extension, Edison steadily expanded his personal technical capabilities and was simultaneously a teacher and mentor to his co-workers.

Although difficult to distinguish from native talent, a separate advantage that Edison brought to his task was his experience. By tackling the most complex electrical problems of the day, in multiple telegraphy and telephony, Edison acquired the most advanced knowledge of the most advanced technology of his time. To be sure, some aspects of electrical science then emerging were beyond the capacity of the unschooled, unmathematical Edison to comprehend. But these were as yet of little or no technical importance. Impressive as Francis Upton's academic credentials were, for example, he exhibited no advantage over Edison in the analysis and solution of the problems of the electric light. Edison's ability to keep up with the new knowledge and theory of electrical technology was to be quickly outrun by the events that he himself set in motion, but this does not change the fact that in 1878 he was as familiar with the basic information underlying applications of electricity as any man living and was well equipped to build on that familiarity in new directions.

Out of this talent and knowledge came yet another indispensable characteristic that Edison brought to his work: confidence. To describe his attitude as confident is, of course, to understate the matter—cocky would perhaps be more to the point. This cockiness was evident from Edison's first claims of a "big bonanza" early in the fall of 1878. He ex-

pressed his confidence to all who cared to listen, and, by this time, there
were many who felt they could not afford to ignore the wizard's promises
and boasts. More than anything else, it was this complete faith in himself
that sustained Edison through the difficulties posed by his initial misper-
ception of the electric light, and he was, in fact, very wrong in much of
his perception of the task ahead of him. He was not bluffing when he
promised to have a complete system in "six weeks or so," just badly mistaken. For more than a year afterwards, enormous self-confidence enabled him to persist in the face of setbacks and disappointments.

The significant extent to which this faith was shared not only by the
companions at Menlo Park but also the watchful and wary men of Wall
Street accounted for the impetus given to the electric light in the fall of
1878. The greatest advantage Edison had over all rivals was this trust
and the men and resources into which it was parlayed. But this advantage was double-edged. The acceptance and commitment of other men's
wealth, as much as his own boastful words, forced Edison to persevere
down ever more unfamiliar and uncharted paths. A certain desperation
undoubtedly crept at times into the spirit at Menlo Park. At these times,
self-confidence had to be bolstered by a sense of obligation and even des-
tiny. Menlo Park itself—the buildings, the equipment, the talented craftsmen employed there—was testimony to the faith of other men and to
Edison's extraordinary ability both to deliver on his promises and to extend ever wider the horizons of his action. With the resources at his
command, he could combine speculations, guesses, and clever tricks to
his heart's content, and then proceed systematically to test the results.
The opportunity to try out ideas, uninhibited by concerns about man-
power or materials and with complete reliance in the skills and talents of
the craftsmen around him, gave Edison a capability possessed by no inventor in history before him. This expansive approach should not be confused with the old image of the ill-directed ransacking of nature's storehouse for lamp filament material or other pejorative parts of the legend
of Edison the empirical experimenter. There were indeed areas of research in which Edison had to work blindly, but this was hardly a special
characteristic of his work nor unexpected in any ambitious technical
enterprise. Edison's resources, however, allowed him to explore blind
alleys as well as to exploit inspirations with an efficiency and speed that
demonstrated for all the real virtues of Menlo Park.

The capital that could be mustered by his reputation was indispensable
also in the fulfillment of the Edison promise at Pearl Street. To be sure,
the caution and conservatism of Wall Street asserted themselves sufficiently to force Edison to put much of his own money into various manufacturing enterprises for the Edison system. But this detracted little
from his ability to summon resources in amounts rarely placed in an inventor's hands. If, in hindsight, Edison's vision seemed wonderfully com-

plete, it was due as much to this command over the means to make the vision whole as to an approach particularly suited to the creation of systems. Indeed, for most of Edison's career, achievements of technical insight are difficult to distinguish from achievements owing more to the great resources he was able to bring to bear on the problems at hand. His "invention factories" were so successful in both concept and execution that the personal element in the inventive act became less distinct as time went on. For this reason, the claim that Edison's greatest invention was the routinization of invention itself has much merit, even if it does not follow that he set the pattern for the corporate research and development laboratories of the twentieth century.

However, invention is not routine. Certainly, great inventions are not, and the incandescent electric light was one of the greatest in history. This is true, in part for the enormous impact it had on our modern technological culture, because it was, as has been pointed out, the indisputable starting point for the creation of the electric power systems that move, control, and inform our life and work. But the greatness of the electric light as an invention does not rest on that alone. The electric light—the lamp and all the power, distribution, and control technology that went with it—was the product of an extraordinary feat of creativity. It is in this creativity that we must seek to understand the ultimate meaning of invention. Other realms in our culture, such as art, music, literature, and science (to name only the most obvious), are commonly scrutinized for evidence about the processes of creation. The study of a painting, a symphony, a poem, or a theory in great detail and with careful attention to the psychological, social, and cultural conditions surrounding its emergence is a common and accepted activity. Such study, it is thought, will bring us closer to the creator and the achievement, will foster our understanding of the values of human culture, and will, just possibly, allow each of us to enhance the expressions of creativity in our own lives. In the realm of technology, however, such scrutiny is rare indeed.

The reasons for this neglect are not difficult to discern. Unlike the products of other creative endeavors, those of the inventor are judged ultimately by one standard alone—the market. That this was a lesson Edison learned early in life is a familiar feature of his biographies. That it was a lesson that guided the work on the electric light is not so common a part of the stories of that invention, but it is obvious from the account here. Immersed in the most exciting creative work of their lives, the men at Menlo Park were moved by many external forces. Besides the overwhelming influence of the market, there were the agitations of investors, the badgering of rivals, the obstructions of bureaucrats and the skepticism of scientists. The rural tranquility of the New Jersey countryside was no shield against the eagerness and anxiety of the world at large. Therefore, the creation of a great new thing was carried out in the

atmosphere typical of technological innovation, in which the act of creation seems overwhelmed by the circumstances and expectations of a
utilitarian world.

Nonetheless, the creative act was there, and it is our responsibility to
seek it out and understand it. That Edison and his companions carried
out their work in a worldly spirit should not blind us to the fact that the
work was a demonstration of the capacities of individuals to envision possibilities inherent in the world around them and to realize those possibilities through persistence, experience, and imagination. We will not and
cannot understand the true origins of novelty—in technology or in any
other part of our lives—unless we acknowledge the personal, human,
creative impulse at its root.

BIBLIOGRAPHICAL
NOTE

The various sources used in this study, and the methods of dealing with them, warrant description and explanation. Such a discussion will also help the reader to understand remarks and references in the endnotes listed for each chapter.

The study relied primarily on the large body of material in the archive at the Edison National Historic Site in West Orange, New Jersey. This extraordinarily complete collection of the inventor's records made it possible to examine the creative process of invention in great detail. The most important sources were laboratory notes and drawings made by Edison and his staff. Because of their special importance, and because such materials have relatively rarely been used for historical analysis, these laboratory records receive the most attention. Correspondence was also of great value, and the varied character of the collection of letters, telegrams, etc. at West Orange requires consideration in some detail.

Somewhat less detail is given in descriptions of other sorts of documents that helped to shed light on Edison's work, but some attempt has been made to evaluate them and to provide the information necessary to gain access to them in the collection at West Orange. Included are discussions of miscellaneous items such as agreements and financial records, collections of clippings from newspapers and journals, primary printed material such as company publications, patents, and litigation records, and reminiscences, both manuscript and published.

Other collections that proved of some value are also discussed. These are the Edisonia Collection in the Tannahill Library at the Edison Institute in Dearborn, Michigan, the Records of the Patent Office at the National Archives, and the Hammer Collection at the Smithsonian Institution.

Notebooks

Laboratory notes and drawings have not been widely used in histories of technology. Often, these sources were simply not preserved, forcing historians to rely on published papers and patents. Even when available, they have generally served only to supplement more traditional documents. A detailed discussion of the Edison laboratory records can illuminate the problems encountered in the use of such records and also address questions raised by the work of others who made previous use of this material in looking at Edison's labors on the electric light.

The complexity of the records makes it best to begin by considering how laboratory notebooks were kept by Edison and his assistants. Fortunately, testimony from patent interference proceedings sometimes touched on this question and helps to reconstruct the system of record-keeping. Testifying in 1880, Edison summarized the evolution of his record-keeping system before he began to work on electric lighting:

> I commenced in a small way, some time in 1873, to have my sketches witnessed, but very few sketches were made compared to those made in 1877, when I commenced the practice of placing note books over all my laboratory, with orders to my assistants to draw out and sign every experiment; previous to 1877 drawings were made on all sorts of scraps of paper and thrown in a drawer.[1]

The notebooks Edison began using in 1877 were 9×11-inch tablets from which pertinent sheets were later removed and reassembled according to the invention to which they referred. In this way Edison provided a record of his experiments for purposes of patent litigation. This system was in use in September 1878 when Edison began his electric light experiments.

At that time he also began to have his bookkeeper, William Carman, copy important notes and drawings into ledgers. The earliest notes on the electric light, from September and October 1878, were copied into "Experimental Researches," Volume 5 (Cat. 997), and the corresponding volume page number was written on each sheet of the original notes and drawings. Originals and other loose pages related to electric light experiments were then grouped together as Notebook 16, each sheet being given a unique page number. The gathering of sheets taken out of tablets became unwieldly when a large number of them accumulated from the many light experiments, and Edison modified his system. In late November he began to use standard 6×9-inch, hardbound notebooks of about 280 pages each. These were numbered for identification and left intact as a permanent record. Existing notebooks date from 1878 to 1882 and number from 1 to 249. Of these, seventy-four are missing, but a few have damaged labels with undecipherable numbers, if numbers once existed. The first missing number is 35, and, of the first fifty notebooks, only five are missing (see Table 1). Thus, the earliest records of electric light experiments are remarkably complete.

The variety of the record-keeping systems being used in the fall of 1878 confused one scholar, who speculated that "since the surviving notebooks from September 1878 indicate no success with subdivision . . . they are incomplete, and it is certainly possible that a book with such an entry has been lost."[2] While items may be missing, they were not ones Edison considered important to his conception of the problem. The most important notes from this time were copied by Carman into a ledger volume within days of being made. They were concerned largely with the design of regulators to control the temperature of the burner, the subject also of the earliest electric light caveats copied into another ledger. Unfortunately, the reminiscences of Edison and his associates have at times led researchers to look for what they expected to find and not what is actually in the record.

Other problems occur in the attribution of dates. According to John Kruesi, superintendent of the laboratory machine shop:

Table 1

Numbers of Notebooks Missing from Archive at Edison National Historic Site

35	69	101	162	181	202	221	246
43	81	109	163	182	205	222	247
44	90	118	164	183	207	232	248
49	91	122	166	185	208	234	
54	92	141	169	193	209	237	
61	93	144	170	194	211	239	
62	94	147	173	195	216	241	
63	97	154	175	196	217	242	
64	98	156	178	199	218	243	
65	99	159	180	200	219		

Three notebooks have labels but unreadable numbers.

Four notebooks have no labels.

The original cover of one notebook, part of the Batchelor Collection, had been replaced, and museum records do not indicate that it had a label.

Two notebooks used by Francis Upton for his literature search are unlabeled.

Notebook 36 is in the Tannahill Library at the Edison Institute.

Other Menlo Park notebooks may be in the Hammer Collection at the Smithsonian Institution.

> . . . the order was in general to date all sketches and all drawings that were made on the day they were made, or in case it was neglected . . . to put down the date of the day when they were dated.[3]

Unfortunately, while this practice was adhered to during the early work on the electric light, Edison and his associates often neglected to date later entries. This seems to have been due to both the pressure of work and the making of notebook entries by more members of the staff.

There are a number of ways to solve dating problems. For example, when, as often happened, two members of the staff made entries in separate books to record the same experiments, a missing date in one book can be deduced if the entry in the other book is dated. Other undated entries can be traced to a certain period if they are preceded and followed by dated entries. Finally, it is possible to ascertain at least the earliest probable date that a particular book was used because the covers of the hardbound books are not all the same. Since books of the same cover color appear to have been bought together, grouping according to color tends to place the books within distinct periods (see Table 2).

Two notebooks that played important roles in major biographies of Edison help to demonstrate dating problems. In his *A Streak of Luck,* Robert Conot claims that an entry in Notebook 190 proves that Edison turned to experiments with carbon filaments in October 1879 because of a *Scientific American* article of July 12, 1879, about which Edison noted:

> In Scientific Amn July 12/79 I notice piece taken from Newcastle on Tyne—Chemical Society's Journal Speaks of Swans having lamp on incandescent principle with *"Cylinder"* carbon = get original paper—"Carbon pencil" spoken of[4]

While this is proof that Edison did read the article and probably tried to get the original Newcastle Chemical Society article, it does not prove he had read it by October 1879. The undated notebook in which Edison's notation appears is one of seven probably begun about January 10, 1881, in preparation for a proposed "Practical treatise on Elec. . . . for Publication in Book Form."[5] These seven, Notebooks 184, 186, 187, 188, 189, 190, and 191, all contain notes on electricity and gas gathered from various published sources. Some also contain draft essays. Several, including Notebook 190, are also mentioned in a memorandum sent to the New York office in February 1881, apparently related to the proposed treatise.[6] Furthermore, these books all have the same cover color and were probably not used until 1880; the earliest dated entry in any book with this cover is January 1880, and most are from June or later. Not only does the evidence about dates weaken Conot's argument, but so does a careful reading of the entry. Edison underlined the word "*Cylinder*" not the word "carbon," indicating that his likely concern was the shape of the carbon used, not the substance itself. Edison was probably underlining the difference between his burner and Swan's.

Notebook 184, also mentioned above as one used in preparation for the proposed treatise, was cited by Matthew Josephson in his 1959 Edison biography, to support the inventor's claim that he began his electric light project by studying the gas industry to determine the relative merits of "Electricity versus Gas as a General Illuminant."[7] Also, in the authorized Dyer and Martin biography of 1910, Edison is quoted:

> On my return home [from William Wallace's shop in September 1878] I started my usual course of collecting every kind of data about gas; bought all the transactions of the gas-engineering societies, etc., all the back volumes of gas journals, etc. Having obtained all the data, and investigated gas-jet distribution in New York by actual observations, I made up my mind that the problem of the subdivision of the electric current could be solved and made commercial.[8]

This account has been repeated by all subsequent Edison biographers, but the notebooks and other documents point to a quite different time for Edison's investigation of the gas industry. While Edison claimed from the beginning that he would replace gas lighting with electric lighting, there is no evidence of any extensive investigation of the gas industry until the spring of 1879. Previous to this there appear only occasional notes on the gas industry, apparently the result of scattered newspaper and journal articles that came to Edison's attention. He did not order books on the gas industry until November 1878 and journals were not received until 1879. The systematic study of gas lighting in New York did not occur until 1880 when surveys were made of the Pearl Street central station district and tests were conducted of the gas lights at Sigmund Bergmann's shop in New York.[9]

As evidenced in the foregoing discussion, the dating of notebooks can be complicated. It requires a certain degree of connoisseurship in treating notebooks as artifacts to be "read" much like archaeological relics. It also requires a careful comparison between dated and undated notebook entries and with other materials. Among these other materials are six pocket-sized notebooks (PN-80-04-09, PN-80-05-03, PN-80-06-08, PN-80-07-14, PN-80-09-23, and PN-81-01-19) and

Table 2

Notebook Dating by Cover Color

Cover Color	Label Numbers	Nov 1878–Apr 1879	May 1879–Dec 1879	Jan 1880–Mar 1880	Apr 1880–Jun 1880	Jul 1880–Sep 1880	Oct 1880–Dec 1880	Jan 1881–Mar 1881	Apr 1881–Dec 1881	Jan 1882–Aug 1882	Oct 1882–Dec 1882	Undated
Blue-green	1–47, 192	36	4	1								2
Dark red	48–96		12	9	2		2					5
Blue-green and red	67–73			5								1
Light blue-black	100–125				6	10	4					1
Green and orange	126–191			2	2	12	11	7	1		3	7
Dark blue-black	197–249							12	11	6		1

two standard-sized notebooks (N-80-03-14 and N-80-07-10) used by Charles P. Mott, a member of the Menlo Park office staff, to record the daily activities of the laboratory from March 1880 to March 1881. Mott often referred to specific notebooks and named workers on specific projects, allowing both names and dates to be attached to otherwise unidentifiable entries.

Besides the Mott journals, there are other ways of determining authorship. Again a certain connoisseurship is required to recognize distinctive handwriting and drawing characteristics. The frequency of entries by Edison, Charles Batchelor, and Francis Upton allows the careful researcher to become familiar with their styles and to distinguish among them with relative ease, but some entries are not so easily identified. Most of the notes from 1878 and 1879 are by these three men. Entries by other members of the laboratory staff are less frequent and, if a writer neglected to sign an entry, identifying the author can be difficult. Again, it is sometimes possible to recognize handwriting, especially if the hand appears repeatedly or in other documents that can be used by comparison. In 1880 many more individuals made entries in lab records, and it is more difficult to distinguish between them, but the Mott journal helps determine the authorship of some entries, which allows the researcher to become familiar with the different writing and drawing styles of staff members.

Another problem presented by the notebooks is the rough character of many of the entries. Thus, while notes are complete on some experiments such as those performed on various metals in January 1879 (see Chapter 2), Edison often made only rough sketches while he explained his ideas orally to his assistants.[10] This resulted in many scribbled drawings with no explicit indication of what they represent. The records kept by other members of the staff also varied in completeness. Francis Upton, perhaps because of his university training, tended to be more thorough than others in recording details of the experiments he performed, but he also produced page after page of unexplained calculations. One important piece of evidence often missing from the notebooks is the reasoning behind experiments or a chosen direction of research. For example, in the case of generator design, the notebooks do not explain why the ring armature and large field magnets were adopted. However, on the basis of Edison's known familiarity with Faraday and notebook sketches of fields of force, it is possible to infer some of his thinking. Or, again, the reasons for the critical decision to turn to carbon in October 1879 are never openly stated and must be deduced from a newspaper account, Edison's requirements for his filament, the notebook entries themselves, and a general knowledge of Edison's work with carbon for other purposes.

Another aid to understanding cryptic notebook entries are the patents and caveats that describe Edison's specific devices. Drafts of patent applications and caveats are found in the notebooks, often giving more detail than the official documents filed with the patent office. British patents are also useful because they allowed the inventor to be more inclusive in his claims and to describe a whole system rather than only individual components.

A word should be said about the archival arrangements of the notebooks. All of the hardbound, standard-sized notebooks used by Edison at Menlo Park and in his later laboratories were assigned "N" numbers by the archive, based on the earliest date in each book. For example, the earliest date in N-78-11-22 is November 22, 1878. Unfortunately, the dates are often incorrect (sometimes due to wrong guesses), so N numbers should never be used for dating notebooks. However, these numbers are important for retrieval because they designate shelf location. For this reason, the endnotes following this section cite N numbers rather than Menlo Park Laboratory numbers.

Other notes and drawings related to electric lighting are found in four other groups of notebook materials. Most are in "Volume 16," Edison's label for the collection of loose pages on subjects pertaining to the electric light. Each page also has a unique page number. Some of this material, from September and October 1878, was also copied into "Experimental Researches," Volume 1, Cat. 994. Some electric light draft caveats were written into "Experimental Researches," Volume 5, Cat. 997. A small group of loose pages of notes and drawings, not included in Volume 16, has been filed, year by year, in a collection called "Miscellaneous Notes and Drawings." Finally, there are Cat. 1146 through 1153, eight scrapbooks in which notes and drawings have been pasted, containing materials from 1878 to 1883 related primarily to the electric light. Most of these notes and drawings were made by Edison in the preparation of patent applications and caveats.

Correspondence

The most fruitful way to discuss the correspondence is first according to archival organization and then the type of material available. Loose correspondence, most of it incoming, is organized in a series known as the "Document File," which also contains smaller quantities of other materials such as agreements and account records. It is arranged by year and, within each year, by main subject categories related to inventions. These subject categories are further subdivided by companies or foreign countries. There is also, for each year, a file known as "Edison, T. A.—General," which contains miscellaneous correspondence and correspondence that deals with more than one subject. The Document File is being reorganized in preparation for a microfilm edition of the Edison Papers and a unique identifying number is being assigned to each folder. Reorganization is complete for all years up to 1886. The following are titles of folder groups relevant to the electric light and a brief summary of the type of information they contain.

Folders found in all years from 1878 through 1882:

Electric Light—General
Inquiries about Edison's work on the light. Many of the letters deal with technical questions, and Edison often wrote notes in the margins to be used in answering the correspondent. These marginal notes are often the only record of an answer. Other letters relate to business arrangements for the light. Of particular note are letters by George F. Barker, professor of physics at the University of Pennsylvania.

Electric Light—Edison Electric Light Company
Material related to the formation and operation of the company. Especially important for 1878 and 1879 are letters by Grosvenor P. Lowrey, the Western Union lawyer instrumental in organizing the company. An 1882 folder contains reports to Edison from Sherbourne V. Eaton, vice-president and chief operating officer, and the most important member of the company between 1880 and 1882. Other 1882 folders hold technical correspondence from Charles L. Clarke, the company's chief engineer, and papers dealing with manufacturing and service companies established for the commercial introduction of the light (filed in the Light Company folder or in individual folders for each company).

Electric Light—Foreign
Items pertaining to foreign business arrangements for the light. Especially important are 1878 and 1879 letters by Lowrey, and, in all years, letters by George E. Gouraud, Edison's agent in England; Theodore Puskas and Joshua F. Bailey, his agents in France; Egisto Fabbri, the Drexel, Morgan & Co. partner; and Edward H. Johnson, who was in England for a time in connection with Edison's light and telephone and who was in charge of the establishment of the Holborn Viaduct station. For 1880–1882, the foreign folder is further subdivided into Europe and United Kingdom. Beginning in 1880, there is also an Edison Electric Light Company of Europe folder.

Edison, T. A.—General
Letters dealing with more than one subject, many from Barker, Gouraud, Johnson, and Puskas.

Patents
Material related to Edison's American and foreign patents. Correspondence is mostly from Edison's chief patent attorney, Lemuel W. Serrell, with some letters from other attorneys or Edison associates. The Electric Light-Foreign-United Kingdom folders also contain some papers on Edison's British patents.

Menlo Park Laboratory
Letters pertaining to purchase of supplies and equipment for the laboratory. An 1881 folder contains reports to Edison by Thomas Logan, superintendent

of the scaled-down laboratory. Purchasing items are also found in other places. See "Edison, T. A.—Bills and Receipts," below under "Other Documents," and Invoice Books, Cat. 1161–1162.

Edison, T. A.—Book and Journal Orders
Letters pertaining to the purchase of books and journals by Edison, many related to his work on the electric light.

Folders on electric light manufacturing and service companies, 1880–1882:

Electric Light-Bergmann & Company (1881–1882)
This company was formed by Edison, Sigmund Bergmann, and Edward H. Johnson to manufacture fixtures and small components for the electric light system. Most of the material deals with orders. Bergmann previously had his own company, and material related to this company are in folder titled "Bergmann, S."

Electric Light—Edison Lamp Company (1880–1882)
Much correspondence from Francis Upton, superintendent of the lamp company, on lamp experiments and production problems. Also material related to orders and operation of the company.

Electric Light—Edison Machine Works (1881–1882)
Material related to production of generators and meters, orders, and company operations. Of special note is the 1882 folder for the Testing Room, which contains correspondence from William S. Andrews, head of the Room, about experiments and tests of meters and dynamos. (Clarke correspondence in the Edison Electric Light Company folder of 1882 (listed above) has similar material about experiments and tests.)

Electric Light—Electric Tube Company (1881–1882)
Primarily correspondence from John Kruesi, superintendent of the company, dealing with orders and operation of the company.

Electric Light—Edison Electric Illuminating Company of New York (1880–1882)
This company was formed in December 1880 to install the first central station in New York. Folders contain some information about the Pearl Street station.

Electric Light—Edison Company for Isolated Lighting (1881–1882)
This company was formed to install isolated lighting plants. In 1882, there are reports sent to Edison containing much technical information excerpted from the company's defect books, which were used to record complaints.

Other important folders:

Exhibitions—Paris Electrical Exhibition (1880–1881)
Material on 1881 electrical exhibition in France, including extensive correspondence by Batchelor, Bailey, and Otto Moses, one of Edison's chemists from Menlo Park who assisted Batchelor in running the exhibition. Some of the correspondence also relates to the establishment of European lighting companies and the promotion and manufacture of the light in Europe.

Mining—Platinum Search (1879)
Correspondence concerning Edison's search for a source of platinum for the
burner in his lamp. Many letters have marginal notes by Edison.

Railroad—Electric (1880–1882)
Material related to Edison's electric railroad and the development of electric
motors. (See also "Mining—General" and Cat. 2174.)

Prior to 1882 letterbooks were used sporadically to record outgoing correspondence. Often, the only record of a response is a marginal notation by Edison
on the incoming letter in the Document File. That file also contains occasional
drafts of outgoing letters. Fortunately, the letterbooks, assigned "LB" numbers, are well indexed so that it is easy to find correspondence sent to important
individuals mentioned in the discussion of electric light materials stored in the
Document File. Of special note are letters written in 1882 to Johnson and
Batchelor, which detail much of the work on the development of the system.
There are fifteen letterbooks (LB-003 to LB-017) dealing with the period 1878–
1882 in the main collection. There is also a letterbook (Cat. 290) kept by Samuel
Insull, Edison's secretary, between January 1882 and November 1883 with much
material on the electric light. Another book (E-4395) contains copies of foreign
cable messages from March 1881 to July 1883, most of which deal with the electric light.

Other Documents

A few other types of documents provide information about the development of
Edison's electric light system. These include agreements, financial records,
memoranda, and reports. Agreements give useful information about Edison's relationship to the various electric light companies. Financial records tell about
Edison's finances, expenditures on experiments, and the operation of the companies, particularly the manufacturing companies. Memoranda and reports contain material of varying importance about Edison's research and the various
companies.

Agreements are found in the Document File in folders related to the companies and in the foreign folders. Another collection, the Harry F. Miller File,
consists primarily of agreements, including some related to the electric light. A
file of photocopies from outside collections contains a few other agreements.
This file is currently being organized by the Thomas A. Edison Papers.

Financial records are stored in the Document File in folders for particular
companies, and in special folders, including:

Edison, T. A.—Accounts
Rough accounts and trial balances, with some summary statements related to
Edison's personal finances and the operation of the laboratory.

Edison, T. A.—Bills and Receipts
Loose bills and receipts. Most of the bills from this period are found in invoice
books (see below). These loose bills are primarily for purchases for the laboratory. Company bills are in folders for individual companies. Bills related to book

and journal orders are in the invoice books and in "Edison, T. A.—Book and Journal Orders" folders. Bills for patent expenses are in the Patents folders.

The most important account books are the 1878–1882 ledgers (Cat. 1184–1187) and two uncatalogued books covering the periods March–December 1880 and January 1881–December 1887. The 1878–1880 books include accounts for various experiments. Also important is a book (Cat. 1224) with weekly statements of money spent on electric lighting from October 1878 to January 1880. These statements were written on individual sheets and pasted in. Invoice books (Cat. 1161–1162) contain 1878–1882 bills annotated to indicate the account they were paid from, sometimes for specific experiments and sometimes for general laboratory expenses. These accounts correspond to those set up in the account books.

Memoranda, reports, and occasional documents such as circulars are found in the Document File in appropriate correspondence folders as described above.

Clippings

Two sets of clippings from newspapers and technical journals give useful information about the work of Edison and his rivals on electric lighting. Newspaper clippings, in particular, tell much about Edison's work that is not easily discerned from notebooks and other materials. For example, they help to better understand his work on generators. Edison, who often made announcements about his work through the press, began his work on electric lighting with the assumption that he could use William Wallace's generator (*New York Sun,* September 16, 1878). By December he had begun work in earnest on generators and soon announced that Wallace's generator was inadequate for his purposes and that he was designing his own generator (*New York Sun,* December 19, 1878). This tends to refute scholars who have argued that Edison envisioned "the essence of components and their interactions before he began intensive research and experimentation."[11] Newspaper interviews considered in conjunction with the notebooks show that Edison's decision to design his own generator resulted from his experimental research and not from earlier theoretical analysis of the "essence of components and their interaction." The clippings also help in reconstructing the way the laboratory operated (see Chapter 2).

The set of clippings that most elucidates Edison's work is a volume (Cat. 1241) put together by Charles Batchelor, which contains primarily newspaper clippings from 1878 to 1881. Also useful are Batchelor clipping books (Cat. 1242–1243) from 1881 to 1882. All the Batchelor clipping books are found in the Charles Batchelor Collection (see below).

Another set of clippings was kept in the laboratory. Menlo Park Scrapbooks, Volumes 1–40, are a set of fifty-seven technical scrapbooks (many subnumbered) and an index volume. They are the first in a series begun by William Carman and Francis Upton in 1878–1879 and augmented and updated by the laboratory staff. The first forty volumes deal mostly with technologies and subsequent volumes added through 1889 with scientific matters. Approximately 150 scrapbooks are extant of a series that may have numbered over 200 books at one time. Most of the clippings are from technical and scientific journals, although some are from popular magazines and newspapers. The scrapbooks on electric lighting include Cat. 1011–1019, labeled "Electric Light," which contain many clippings dealing directly with Edison's work; Cat. 1025–1027, labeled "Magneto Electric Generators"; Cat. 1051, labeled "Electric Lamp"; and Cat. 1052, labeled "Radiometer and Vacuum Pump." A few other volumes contain material about electricity and magnetism useful for understanding the state of the art at that time. (Also see Cat. 1002, "Combustion of Coal; Theoretical Heat from Boilers and Steam Engine Cost.")

Primary Printed Matter

Various printed materials were useful in the study of the development of the electric light. These include the Edison Electric Light Company *Bulletins,* company advertising brochures and catalogs, issued patents, and patent litigation records.

The Edison Electric Light Company *Bulletins* were begun in January 1882 to answer questions of the company's agents and were later also sent to company stockholders. Twenty-one were published between January 1882 and December 1883. They include information about the various electric light companies and the annual reports of the Edison Electric Light Company, the Edison Illuminating Company of New York, and the Edison Company for Isolated Lighting. They are a particularly good source following the progress being made between January and September 1882 on the installation of the Pearl Street central station in New York. The *Bulletins* also give accounts of other central and isolated plants in the United States and abroad and items of interest about the Edison and rival systems of electric lighting, gas companies, and electric lighting in general, including reprints of newspaper and journal articles. Bound sets of these volumes are located at West Orange.

The various Edison electric light companies published advertising circulars, catalogs, and brochures for their agents to use in selling the system. These have been gathered into a collection of "Primary Printed Material for Edison Companies." A similar collection contains material from non-Edison companies but little material from the 1880s. These collections have been indexed by the Thomas A. Edison Papers. There is also a bound volume of some of the Edison companies' pamphlets that is not yet catalogued.

All of Edison's patents have been bound into seven volumes in order of patent number. Most of the electric light patents from 1878 to 1882 are in Volumes 1–3. The Edison Electric Light Company also kept records of Edison's patent applications from 1878 to 1885. The claims and drawings of these patent applications were copied into uncatalogued volumes E-2534–2535, E-2537–2538, and E-4398–4400.

Bound volumes of patent litigation records, including some electric light cases, are in the library at the Edison National Historic Site. An inventory has been prepared by the Thomas A. Edison Papers. Of particular note are *Edison Electric Light Co. vs. United States Electric Lighting Co.* (7 volumes) on U.S. Patent 223,898 for the carbon filament, and the testimony from patent interferences, *Sawyer and Man vs. Edison* (filament), *Keith vs. Edison vs. Brush* (dynamo electric machines), and *Mather vs. Edison vs. Scribner* (dynamo electric machines). Exhibits from patent litigation often provide the only extant copies of some documents. The testimony is particularly useful for fleshing out what was going on in the laboratory but must be corroborated by other materials to substantiate its validity. Robert Conot, in his biography of Edison, noted that the inventor was a poor witness when asked to recall the sequence of events and that "they blended and telescoped into each other, and he misplaced even fairly recent happenings by as much as three months."[12] This statement about Edison's testimony in the complicated court cases on the quadruplex telegraph apply also to his testimony in the electric light cases and his still later reminiscences about his work on electric lighting.

Reminiscences

The reminiscences most widely used by other researchers into the development of the electric light have been those of Edison, extensively quoted in the Dyer and Martin official biography, *Edison: His Life and Inventions,* and of Francis Jehl, a laboratory assistant who wrote the three-volume *Menlo Park Reminiscences.* There are also less well-known and shorter reminiscences by Edison and his associates, some of which were written for the Dyer and Martin biography.

Edison also wrote notes and answers to questions for his secretary, William H. Meadowcroft, which were used in the Dyer and Martin biography. Although Meadowcroft is cited as a collaborator in the biography, he was very likely the principal author. Some of this material was also used for a short article Edison wrote for Henry Ford in 1926 for which Meadowcroft probably acted as ghostwriter.[13] In the article, Edison incorrectly remembered investigating the gas industry before he began work on the light, a typical Edison failure as noted above, to recall accurately when and in what order certain events took place. Such a lapse appears even in his legal testimonies, which were usually much closer in time to the actual event than were his reminiscences.

Another problem with the reminiscences of Edison and his associates is the tendency to describe events as they are given in the "official" story even when they are unlikely to have happened that way. An example is the famous "forty-hour" lamp of October 21, 1879. In the article Edison wrote for Ford, he says that the laboratory notebook entry for this date records that a lamp made of cotton thread lasted about forty-five hours. In fact, it lasted thirteen and a half. This same exaggerated story has been told by others who were present at Menlo Park in October.

Among those present was Francis Jehl, who was eighteen when he joined Edison as laboratory assistant. In his *Menlo Park Reminiscences,* Jehl claimed to have been there in November 1878 and even described the arrival of Francis Upton at the laboratory at that time. To the contrary, employment records, including a letter of recommendation from Grosvenor P. Lowrey and Jehl's own testimony in patent interferences, show that he was not there until February 1879. Jehl also gave himself a more prominent role in the early electric light research than is indicated by the laboratory records. He does not figure in the notebooks until September 1879 when he assisted Francis Upton with vacuum pump experiments. Later that year and in 1880, Jehl seems to have been engaged primarily in testing lamps, and, during 1880 and 1881, assisted in the development of the vacuum pump and the chemical meter.

Jehl's *Reminiscences* does have its merits, since parts are based on a diary which internal evidence suggests he kept from February to May 1881 while he was at the Edison Machine Works Testing Room. (Unfortunately, the diary could not be located in the Tannahill Library at the Edison Institute in Dearborn, Michigan.) In his discussion of this period, Jehl quotes directly from his diary. He also used various documents and contemporary articles to refresh his memory so that *Reminiscences* takes on something of the character of a secondary source. Moreover, his discussion of the roles of various members of the laboratory staff is quite useful, and he induced some of them to contribute reminiscences of their own. Of particular note are two that Jehl quoted in their entirety: "Menlo Park in 1880," by Charles L. Clarke, pp. 855–63 and "The Historic Pearl Street New York Edison Station," by Clarke and John W. Lieb, pp. 1041–55.

Other short reminiscences written as part of applications for membership in the Edison Pioneers or otherwise collected by the Pioneers, are stored at the West Orange archive in the Edison Pioneers Biographical Collection. The material is filed under each author's name and is a source of useful information about those who worked with Edison.

The Meadowcraft Box Collection contains reminiscences used in the Dyer and Martin biography, some sent by Edison associates to William H. Meadowcroft when he was compiling material for the biography, and Edison's reminiscences written with Meadowcraft for the biography.

Special collections include a short account by Charles Batchelor of Edison's return from Wyoming and visit to William Wallace's shop in the summer of 1878, and a brief analysis by Francis Upton of Edison's method of invention, which give useful information about Edison's work on electric lighting. (See "Special Collections" below for discussion of Batchelor and Upton Collections containing materials by these principal Edison assistants.)

Special Collections

Two special collections are particularly useful for reserch on Edison's work in electric lighting. These are the Charles Batchelor Collection and the Francis R. Upton Collection.

The Batchelor Collection was given to the Edison National Historic Site between 1957 and 1961 by Batchelor's daughter Emma. Much of the material relates to work on electric lighting between 1878 and 1882. Of particular note are the following items:

1237	Notebook 1180–1899
1239	Notebook 1881–1883
1241	Clippings 1878–1881 (see discussion under clippings above)
1242	Clippings 1881
1243	Clippings 1881–1882
1301–3	Three notebooks containing notes of lamps ordered and tested between September 1880 and October 1882. Entries are numbered 1–1301.

1304   Notebook 1878–1879, 1884–1891, a very important early
            notebook
1308   Order Book 1879, kept by Batchelor and John Kruesi to record
            orders for experimental devices made by the laboratory
            machine shop in 1879. These correlate with notebooks that
            mention order numbers.
1321   Bound volume of Edison's British Patents 1872–1880

Unbound papers in the Batchelor Collection are organized chronically into three
series: general file (correspondence and other documents such as agreements),
technical notes and drawings, and financial records. This collection also has
other materials useful for research on the marketing of the electric light in Europe during 1881–1883. Researchers are helped by the inventory prepared by
the archive staff and by two card indexes to the collection.

   The Upton Collection was given to the Edison National Historic Site in 1963
by Upton's daughter Eleanor Upton via Paul J. Kruesi, son of John Kruesi. Most
of the material in this collection is dated later than 1882 but includes important
materials related to research on the electric light. Of particular note are the following items:

E-6285-1,-2   Notebooks used by Upton in November 1878 for his
               literature search
E-6285-5   Five letters from Upton to Charles Farley, 1878–1880
E-6124   Letters from Upton to his family, 1878–1880
E-6298   Includes letters from Upton to others concerning the
               electric light and a memorandum by Upton of notebook
               references, 1879–1882, dealing with the electric light

Correspondence and memoranda, including items E-6285-5, E-6124, and E-
6298, are organized chronologically.

Other Collections

While the archive at the Edison National Historic Site contains most of the important material related to Edison's work on the electric light, other manuscript
collections were also searched. The most important were the Edisonia Collection in the Tannahill Library at the Edison Institute in Dearborn, Michigan,
the Records of the Patent Office at the National Archives, the Hammer Collection at the Smithsonian Institution, and the Swan Papers in Newcastle-on-Tyne,
England.

   The Edisonia Collection houses the second largest collection of Edison materials, some of which relate to work on the electric light. They proved to be only
supplemental to the material at West Orange. Among the materials are miscellaneous agreements, correspondence, notes and sketches, and company brochures. Edisonia also includes a Pioneers Collection with biographical information, reminiscences, and miscellaneous manuscripts. The Edisonia Collection
was being reorganized at the time it was used, but the reorganization has since
been completed and an inventory is now available.

   The Records of the U.S. Patent Office (National Archives Record Group 241)
contain a number of useful items, particularly:

Patent Application File

Applications related to Edison's allowed patents have been microfilmed, and a set is located at the Edison Papers Office at Rutgers University, New Brunswick, New Jersey. These include the original applications, any changes made, and correspondence between Edison's patent attorneys and the Patent Office. There are also abandoned patent application files but most of these were destroyed with Congressional approval and no early Edison applications were kept.

Interference Case Files

These are not easily accessible, and the National Archives does not have a complete set of all interference proceedings. They are accessible through an inventory prepared by the Archives staff and through the Register of Interferences kept by the Patent Office and now part of Record Group 241. Among the interferences in the collection is one between Edison and Ludwig Boehm, his glassblower in 1879 and 1880, concerning the development of Edison's vacuum pump.

The Hammer Collection at the Smithsonian Institution contains many items relating to the career of William J. Hammer, who became associated with Edison in the development of the electric lighting system in 1880. Of immediate importance to those interested in the work on electric lighting in the period 1880–1882 are some Menlo Park notebooks; Box 12 has material on the installation and operation of the Holborn Viaduct central station in London, and Box 13 has material on the planning of the Pearl Street central station. The collection also contains miscellaneous correspondence and clippings from this period.

NOTES

With few exceptions, all sources cited in the following chapter notes are in the archive at the Edison National Historic Site in West Orange, New Jersey. Other repositories are indicated where appropriate. Location in the archive is designated in various ways: Special Collection (e.g., Batchelor Collection); Document File by year and folder title (e.g., DF 1878, "Electric Light—General"); Edison's notebook Volume 16 and page number (e.g., Notebook, Vol. 16: 6); standard-size "N" notebook and page number (e.g., N-78-11-22: 13–17); pocket "PN" notebook and entry date (e.g., PN-81-01-19: February 5, 1881); Catalog Number and page number (e.g., Cat. 1304: 2); "LB" letterbook and page number (e.g., LB-003: 394); and others that are self-explanatory (e.g., book references). For brevity, Chapter 6 citations from Mott's journals (N-80-30-14 and N-80-07-10) give only Mott's name and journal entry date. Many references to letters in the Document File omit the folder title because correspondence has been indexed by the Thomas A. Edison Papers and is accessible by date, author, and recipient. TAE, of course, is Thomas Alva Edison. For more details on sources, see preceding Bibliographical Note.

Chapter 1. "A Big Bonanza"

1. Matthew Josephson, *Edison, A Biography* (New York: McGraw-Hill, 1959): 175; Robert Conot, *A Streak of Luck* (New York: Seaview Books, 1979): 115.

2. Charles Batchelor, undated memoir (c. 1905) on electric light, Batchelor Collection.

3. Notebook, Vol. 16: 6.

4. Quoted in *The Mail*, September 10, 1878, Batchelor Scrapbook, Cat. 1241.

5. *New York Sun*, October 20, 1878, Cat. 1241.

6. TAE to William Wallace, September 13, 1881.

7. Electric Light Caveat, September 13, 1878, "Experimental Researches," Vol. 5 (Cat. 997): 49–63.

8. Ibid.

9. *New York Sun*, September 16, 1878, Cat. 1241.

10. George F. Barker to TAE, September 16, 1878.

11. Grosvenor P. Lowrey to TAE, September 17, 1878, and Tracy R. Edson to TAE, September 19, 1879.

12. Notebook, Vol. 16: 23–33.

13. TAE to Theodore Puskas, September 22, 1878.

14. George Bliss to TAE, September 24, 1878.

15. Notebook, Vol. 16: 32.

16. Ibid.: 40.

17. Ibid.: 44.

18. Barker to TAE, October 10, 1878.

19. Moses G. Farmer to TAE, October 7, 1878.

20. Notebook, Vol. 16: 127.

21. TAE to Lowrey, October 3, 1878, LB-003: 390.

22. TAE to Puskas, October 5, 1878, LB-003: 394.

23. TAE to George E. Gouraud, October 8, 1878.

24. DF 1878, "Electric Light—General."

25. Ibid. and TAE to Condit, Hanson, and Van Winkle, October 10, 1878, LB-003: 400.

26. Gouraud to TAE, October 16, 1878.

27. Gouraud to TAE, October 24, 1878.

28. *New York Sun*, October 20, 1878, Cat. 1241.

29. Undated cables between TAE and Grosvenor P. Lowrey, c. October 1878.

30. Stockton Griffin to Lowrey, November 1, 1878, LB-003: 467–468.

31. Lowrey to TAE, November 2, 1878.

32. Lowrey to TAE, October 31, 1878.

33. TAE to Lemuel W. Serrell, October 31, 1878, LB-003: 465.

34. DF 1878, "Electric Light—Edison Electric Light Company."

35. Francis Upton to his mother, November 7, 1878, Upton Collection.

36. TAE to Howard Butler, November 12, 1878.

37. TAE to Puskas, November 13, 1878.

38. Ibid.

39. Upton to TAE, November 22, 1878, DF 1878, "Electric Light—Edison Electric Light Company."

40. TAE to William & Rogers, November 29, 1878, LB-004: 20.

Chapter 2.
"The Throes of Invention"

1. The first volume of Francis Jehl, *Menlo Park Reminiscences*, 3 vols., (Dearborn, Michigan: Edison Institute, 1936–1941) is a generally reliable source on the background of the men at Edison's laboratory.

2. See Charles Batchelor Scrapbooks, Cat. 1240 and Cat. 1241.

3. Grosvenor P. Lowrey to TAE, November 25, 1878.

4. Lowrey to Stockton Griffin, December 5, 1878.

5. Lowrey to TAE, December 10, 1878.

6. Lowrey to TAE, December 23, 1878.

7. TAE to Theodore Puskas, January 3, 1879.

8. Lowrey to TAE, January 25, 1879.

9. Josephson makes this point in *Edison*, p. 190 (see Note 1, Chapter 1).

10. N-78-11-22, p. 5.

11. Notebook, Cat. 1304: 13, Batchelor Collection.

12. N-78-11-22, pp. 13–17 and N-78-11-28, pp. 31–33.

13. *New York Sun*, December 19, 1878, Cat. 1241.

14. William Wallace to TAE, December 21, 1878.

15. Wallace and Sons to TAE, December 11, 1878; TAE to George F. Barker, December 19, 1878; Barker to TAE, December 21, 1878; and Henry Morton to TAE, December 26, 1878.

16. Charles H. T. Collis to TAE, December 21, 1878.

17. N-78-11-22, pp. 21.

18. Cat. 1304: 25 and Batchelor to James Adams, January 2, 1879.

19. N-78-12-31, pp. 30–96 and N-78-12-20.1, pp. 70–106.

20. N-78-12-31, pp. 45–47.

21. N-78-12-15.1, p. 121 and N-78-12-20.1, pp. 266–280.

22. N-79-01-19, pp. 27–68.

23. Ibid., 61–62.

24. Lowrey to TAE, January 25, 1879.

25. TAE to Barker, January 22, 1879 and TAE to Morton, January 22, 1879.

26. Notebook, Vol. 16:368.

27. N-79-01-21, p. 41.

28. N-78-12-31, pp. 99–101.

29. Ibid., 101–103.

30. Ibid., 105–163.

31. N-79-02-20.1, pp. 63–67.

32. N-79-02-24.1, pp. 51–87.

33. N-79-02-15.2, pp. 31–161. (British Patent 2402 of 1879 issued June 17, 1879.)

34. Cat. 1304:2.

35. Ibid., 5.

36. N-78-12-20.3, p. 3 and N-78-12-16, pp. 1–16.

37. N-79-02-15.2, p. 105.

38. N-79-02-24.1, pp. 79–81.

39. Francis Upton to his father, February 23, 1879, Upton Collection.

The Search for a Vacuum

1. N-79-01-19, p. 47.

2. National Archives and Records Service. Record Group 241. Interference 7943: Boehm vs. Edison. No. 18, testimony on behalf of Edison, p. 40.

3. Ibid., Interference 7768: Sawyer and Man vs. Edison. No. 35, testimony on behalf of Edison, p. 15.

4. N-78-12-31, p. 97 ff.

5. Interference 7943, p. 40

6. Reinmann & Baetz to TAE, March 25, 26, and 27, 1879.

7. Interference 7943, pp. 37, 43.

8. Ibid., 18–19.

9. Ibid., 43. The article was: W. de la Rue and H. W. Muller, 1878. "Recherches experimentales sur la décharge électrique avec la pile a chlorure d'argent," *Annales de Chimie et de Physique,* 15, pp. 289–231; the pump combination is pictured on p. 294.

10. Interference 7943, No. 19, p. 8.

11. Interference 7768, No. 35, p. 116.

12. Ibid., 70.

13. Interference 7943, No. 18, p. 47.

14. Interference 7768, No. 35, p. 24.

Chapter 3. "Some Difficult Requirements"

1. N-78-11-28, pp. 7–9.

2. N-78-12-15.1, p. 15.

3. N-79-04-03, pp. 27–41.

4. N-79-02-15.2, pp. 135–145.

5. Ibid., 133.

6. Ibid., 147–151.

7. N-78-11-28, pp. 1–5.

8. N-79-03-25, p. 72.

9. See, for example, N-78-12-04.2; N-78-12-11; N-78-12-15.1; N-78-12-28; N-78-12-31; N-79-01-01; N-79-01-19; N-79-02-15.1; and N-79-02-24.1.

10. Cat. 1308:119.

11. Ibid., 121.

12. Ibid., 137.

13. Ibid., 141.

14. Ibid., 143.

15. N-79-02.15.2, pp. 151–153.

16. Ibid., 155.

17. Francis Upton to his father, April 13, 1879, Upton Collection.

18. *New York World,* April 30, 1879, Cat. 1241.

19. Ibid.

20. U.S. Patents 214,636 and 214,637.

21. *New York Herald,* April 27, 1879, Cat. 1241.

22. Ibid.

23. N-79-03-10.1, pp. 51–56; N-79-01-14, pp. 73–81; N-79-04-03, pp. 47–61; N-79-01-21, pp. 129–133 and 141–149; and Cat. 1308 (Menlo Park Laboratory machine shop order book): 133–135, Batchelor Collection.

24. N-79-03-10.1, pp. 37–39.

25. Letters from Reinmann & Baetz, March 25, 26, and 27, 1879.

26. N-78-11-21, p. 119.

27. TAE to J. O. Green, April 18, 1879.

28. Upton to his father, April 4, 1879, Upton Collection.

29. *Encyclopaedia Britannica*, 11th ed., Vol. 6, pp. 856–857.

30. N-78-12-11, pp. 280–282.

31. Cat. 1304, pp. 44–45.

32. N-79-03-25, pp. 1–9.

33. DF 1879, "Mining—Platinum Search"; TAE to Calvin Goddard, May 26, 1879, LB-004, pp. 355–357; and TAE to Scientific Publishing Company, May 26, 1879, LB-004: 358.

34. DF 1879, "Mining—Platinum Search" and TAE to U.S. Minister, St. Petersburg, Russia, July 18, 1879, LB-005: 2.

35. N-79-07-25, p. 3; N-79-07-31, pp. 264–278; and N-79-08-22, pp. 31, 45, and 54.

36. N-79-08-22, pp. 96–103.

37. Ibid., 3–134.

38. Ibid., 132–133 and N-79-09-20, p. 13.

39. N-79-08-22, p. 129.

Carbon and the Incandescent Lamp

1. Arthur A. Bright, *The Electric Lamp Industry* (New York: Macmillan, 1949), pp. 36–42; James Dredge, *Electric Illumination* (London: Engineering, 1882): vol. 1, 572–578; John W. Howell and Henry Schroeder, *History of the Incandescent Lamp* (Schenectady: The Maqua Co., 1927): 29–41.

2. National Archives and Records Service. Record Group 241. Interference 8195: Edison vs. Maxim vs. Swan. No. 73, testimony on behalf of Edison, p. 196.

3. Ibid., Interference 7768: Sawyer and Mann vs. Edison. No. 35, testimony on behalf of Edison, p. 132.

4. Ibid., 134–135.

5. Ibid., 34, 142.

6. Ibid., 144.

7. Ibid., 145–146.

8. Ibid., 49, 51, 58.

9. Ibid., Interference 8195, p. 2.

10. Ibid., Interference 7768, p. 17.

11. Ibid., 136.

12. Ibid., 46.

13. Ibid., 56.

14. Ibid., 155.

15. Aaron Solomon to TAE, July 25, 1879.

16. B. S. Proctor, "The Smoke of an Electric Lamp," *Scientific American,* 41 (July 12, 1879): 20.

17. N-79-02-12, p. 5.

Chapter 4.
The Triumph of Carbon

1. N-79-07-31, pp. 85–91.

2. N-79-08-22, p. 135.

3. DF 1879, "Telephone—Carbon Button Orders."

4. *Engineering,* March 21, 1879, quoted in Jehl, *Reminiscences* 1: 277 (see Note 1, Chapter 2).

5. DF 1879, "Telephone—Foreign—United Kingdom."

6. Ibid. and N-79-01-21, pp. 239–241, 245; N-79-06-12, pp. 73–75; N-79-09-18, pp. 45, 97.

7. N-79-08-22, p. 169.

8. N-79-07-31, pp. 93–97.

9. Ibid., 99–103.

10. Ibid., 105.

11. N-79-08-22, p. 171.

12. N-79-07-31, p. 107.

13. Ibid., 111–115.

14. N-79-08-22, pp. 173–175.

15. N-79-07-31, p. 117.

16. Ibid., 119–257.

17. Francis Upton to his father, November 2, 1879, Upton Collection.

18. Ibid., November 9, 1879.

19. Ibid., November 16, 1879.

20. U.S. Patent 223,898, January 27, 1880.

21. Upton to his father, November 22, 1879, Upton Collection.

22. Telegrams between TAE and George E. Gouraud, December 1, 1879.

23. TAE to Norvin Green, November 4, 1879.

24. N-79-04-03, pp. 174–282.

25. TAE to C. G. Wildreth, November 17, 1879, LB-005:359.

26. Grosvenor P. Lowrey to TAE, November 13, 1879.

27. Stockton L. Griffin to Joshua F. Bailey, December 2, 1879, LB-005:389.

28. Upton to his father, December 7, 1879, Upton Collection.

29. TAE to "Phonos" (cable code for Edward H. Johnson), December 17, 1879.

30. Upton to his father, December 21, 1879, Upton Collection.

31. Eggisto P. Fabbri to TAE, December 26, 1879.

32. Upton to his father, December 28, 1879, Upton Collection.

33. *New York Herald,* December 28, 1879, Cat. 1241.

34. *New York Herald,* January 1, 1880, Cat. 1241.

Chapter 5. Business and Science

1. N-79-12-00, pp. 1–11.

2. Calvin Goddard to TAE, December 27, 1879.

3. TAE to Goddard, December 29, 1879, LB-005:475.

4. Francis Upton to Charles B. Farley, January 25, 1880, Upton Collection.

5. Undated memoir from Upton Collection.

6. See N-80-01-30, N-80-01-31, and N-80-02-08.2.

7. N-80-01-26, pp. 25–29.

8. Ibid., 31–33.

9. Ibid., 70–128.

10. N-80-01-02, pp. 1–3 and 13–15.

11. N-79-06-16.2, pp. 215–220.

12. N-79-01-14, pp. 150–151.

13. N-79-06-12, pp. 98–101.

14. N-79-10-18, pp. 221–222.

15. N-79-06-12, p. 107.

16. N-79-10-18, p. 234.

17. N-80-01-26, p. 45.

18. Ibid., 49.

19. Upton to his father, January 25, 1880, Upton Collection.

20. N-79-08-28, pp. 1–37 and Howell folder in Edison Pioneers Biographical File.

21. Alfred Taylor to TAE, January 3, 1880.

22. G. Haines to TAE, January 6, 1880.

23. Austin Kenny to TAE, January 18, 1880.

24. Letters are in DF 1880, "Electric Light—General."

25. Simon Newcomb to TAE, January 19, 1880.

26. See David Hounshell, "Edison and the Pure Science Ideal in America," *Science* 207 (February 8, 1980):612–617.

27. N-79-03-10.2, pp. 91–103.

28. N-79-02-14, pp. 77–79.

29. N-80-01-02.2, pp. 26–57 and 91 to end.

30. Ibid., 94–168 and N-80-02-16, pp. 1–7.

31. N-80-01-02.2, pp. 59 and 73.

32. Ibid. 63 and 71–79; also see N-80-03-29.

33. N-80-02-16, p. 83.

34. N-80-01-26, pp. 35–43.

35. Report of Cyrus F. Brackett and Charles A. Young, March 27, 1880, DF 1880, "Electric Light— General."

36. Henry A. Rowland and George F. Barker, "On the Efficiency of Edison's Electric Light," *American Journal of Science* 19 (April 1880): 337–339.

37. Draft of letter from TAE to Joseph Medill, April 8, 1880.

38. Ibid.

39. Henry Morton et al., "Some Electrical Measurements of One of Mr. Edison's Horseshoe Lamps," *Scientific American,* 42 (April 17, 1880): 241.

40. N-78-06-12, pp. 173–174 and N-78-11-21, pp. 136–141.

41. N-80-03-15, pp. 207 and 217.

42. Ibid., 209–213.

43. Upton to his father, May 9, 1880, Upton Collection.

44. "The Columbia," *Scientific American,* 42 (May 22, 1880): 326.

The Menlo Park Mystique

1. National Archives and Records Service. Record Group 241. Interference 7943: Boehm vs. Edison. No. 18, testimony on behalf of Edison, p. 40.

2. Francis Upton to his father, April 27, 1879. Upton Collection.

3. Interference 8195: Edison vs. Maxim vs. Swan. No. 73, testimony on behalf of Edison, p. 195.

4. Mary C. Nerney, *Thomas A. Edison, A Modern Olympian* (New York: Harrison Smith and Robert Haas, 1934): 64.

5. Interference 8195. No. 105, record in support of J. W. Swan, p. 558.

6. George Bernard Shaw, *The Irrational Knot* (London: A. Constable, 1905): xi.

7. Interference 8195, No. 105, p. 274 ff.

8. Ibid., 156.

9. Interference 7943, No. 18, p. 38.

10. Nerney, *Edison,* 57.

11. Manuscript, W. H. Meadowcroft typescript of Edison's "Autobiographical Notes," 2:13–14, Edison National Historic Site.

12. Interference 8195, No. 105, p. 29 and Francis Jehl, *Reminiscences* (see Note 1, Chapter 2), 1:285.

13. Upton to his father, March 2, 1879, Upton Collection.

Chapter 6.
A System Complete

1. The Mott journals are notebooks N-80-03-14 and N-80-07-10; his pocket notebooks (draft daily records) are PN-80-04-09, PN-80-05-03, PN-80-06-08, PN-80-07-14, PN-80-09-23, and PN-81-01-19. For brevity, journal citations in the text give only Mott's name and the entry date.

2. Mott: July 15, 1880; also N-80-06-02, pp. 23–31 and N-80-06-28, pp. 37–43.

3. Some of this information is from "Thomas Alva Edison Lighting Timeline," unpublished ms. in Corning Glass Works Corporate Archives.

4. N-80-06-02, pp. 73–77.

5. N-80-10-01, p. 51; N-80-12-13, pp. 3–7; N-80-11-16, pp. 149 and

163; and Mott: December 15, 17, and 21, 1880.

6. N-78-11-21, pp. 136–141 and N-79-06-12, pp. 196–197.

7. N-80-09-11, pp. 17–19.

8. N-80-08-10, pp. 53–77.

9. Charles L. Clarke's report was dated February 7, 1881 and, as he explained later, Edison was so pleased with the results he initially planned to publish it, but commercial considerations intervened. The copy of the report used here was given by Clarke to the library of the American Institute of Electrical Engineers in 1904 and it is now in the Engineering Societies Library, New York.

10. Jehl, *Reminiscences,* 2:876–878 (see Note 1, Chapter 2).

11. Ibid., 2:879–888.

12. TAE to Grosvenor P. Lowrey, July 20, 1880.

13. N-80-06-29, pp. 145–155; N-80-07-16, pp. 3–139; and N-80-07-05, pp. 59–120.

14. Jehl, *Reminiscences,* 2:723–724.

15. G. W. Soren to TAE, November 5, 1880.

16. TAE to Bermeire Magis, October 19, 1880.

17. Tracy R. Edson to TAE, November 20, 1880.

18. Draft prepared by Goddard, December 15, 1880, and draft from TAE to Aldermen, December 18, 1880.

19. *New York Truth,* December 21, 1880.

20. Jehl, *Reminiscences,* 2:770–777.

21. Lizzie Upton to her sister Sadie, December 27, 1880, Upton Collection.

Chapter 7.
Promises Fulfilled

1. PN-81-01-19: February 5, 1881.

2. Ibid., March 10, 1881.

3. Grosvenor P. Lowrey to TAE, December 17, 1880.

4. Payson Jones, *A Power History of the Consolidated Edison System, 1875–1900* (New York: Consolidated Edison Company, 1940): 111–119.

5. Ibid.: 350–351.

6. These agreements are found in DF 1881, "Electric Light—Edison Electric Lamp Company—General."

7. TAE to Edward H. Johnson, November 23, 1881, LB-010:, 331–354.

8. Jehl, *Reminiscences,* 2:848 (see Note 1, Chapter 2).

9. PN-81-01-19: February 18, 1881.

10. John Kruesi to TAE, June 11, 1882.

11. This story is told in Josephson, *Edison,* p. 231 (see Note 1, Chapter 1).

12. *New York Tribune,* August 14, 1882, quoted in Payson Jones, p. 140.

13. TAE to Owen Gill, January 29, 1881, LB-006:874.

14. TAE to L. Prang & Co., February 11, 1881, LB-006:919.

15. S. B. Eaton to TAE, November 23, 1881, and circular of November 11, 1881, issued by the Edison Electric Light Company, DF 1881, "Electric Light—Edison Company for Isolated Lighting."

16. Printed list in DF 1882, "Electric Light—Edison Company for Isolated Lighting."

17. TAE to Louis de Bebian, February 5, 1881, LB-006:902.

18. Edison Electric Light Company, *Bulletin,* No. 4, March 8, 1882, p. 3.

19. Draft agreement dated April 1881, DF 1881, "Bergmann & Company."

20. Jehl, *Reminiscences,* 2:563.

21. Edison Electric Light Company, *Bulletin,* No. 15, December 20, 1882, p. 44.

22. TAE to Charles Batchelor, December 31, 1881, LB-010:489–495.

23. Charles L. Clarke to TAE, November 11, 1881.

24. William Hammer to Frank W. Smith, August 31, 1932, Edison Pioneers Biographical File.

25. TAE to Edison Electric Light Company, December 2, 1880, LB-006:610.

26. N-80-08-13.

27. N-80-11-25, pp. 27–141 and N-80-08-13, pp. 49–255.

28. Edward H. Johnson, "Edison Electric Light Stock considered as a speculative holding for the ensuing quarter," September 15, 1881, DF 1881, "Electric Light—Edison Electric Light Company."

29. Letter to the stockholders of the Edison Electric Light Company, April 19, 1881, DF 1881, "Electric Light—Edison Electric Light Company."

30. Clarke to TAE, June 3, 1882.

31. Clarke to TAE, August 1, 1882.

32. Draft agreement dated April 1881, DF 1881, "Electric Light—Edison Machine Works."

33. Samuel Insull to Batchelor, September 28, 1882.

34. N-81-04-06, pp. 37–41.

35. TAE to S. B. Eaton, September 13, 1881.

36. Edison Electric Light Company, *Bulletin,* No. 9, May 15, 1882:7–11.

37. TAE to Johnson, December 31, 1881.

Chapter 8. Afterword

1. Conot, *Streak of Luck,* p. 471 (see Note 1, Chapter 1).

2. Abbott Payson Usher, *A History of Mechanical Inventions* (Cambridge, Mass.: Harvard University Press, 1929): 74–77.

3. Harold C. Passer, *The Electrical Manufacturers, 1875–1900* (Cambridge, Mass.: Harvard University Press, 1953) and Arthur A. Bright, *The Electric-Lamp Industry: Technological Change and Economic Development from 1800 to 1947* (New York: Macmillan, 1949).

4. Christopher S. Derganc, "Thomas Edison and His Electric Lighting System," *IEEE Spectrum,* 16 (February 1879): 50–59; David A. Hounshell, "Edison and the Pure Science Ideal in America," *Science,* 207 (February 8, 1980): 612–617; and George Wise, "Swan's Way: A Study in Style," *IEEE Spectrum,* 19 (April 1982): 66–70.

5. The most important of Thomas Hughes' writings on Edison are *Thomas Edison: Professional Inventor* (London: The Science Museum, 1976); "Edison's Method," in *Technology at the Turning Point,* ed. William B. Pickett (San Francisco: San Francisco Press, 1977): 5–22; "Inventors: The Problems They Choose, the Ideas They Have, and the Inventions They Make," in *Technological Innovation: A Critical Review of Current Knowledge,* ed. P. Kelly and M. Kranzberg (San Francisco: San Francisco Press, 1978): 168–182; "The Electrification of America: The Systems Builders," *Technology and Culture* 20 (April 1979): 124–161; and *Networks of Power: Electrification in Western Society, 1880–1930* (Baltimore: Johns Hopkins Press, 1983). In this last work, only the second

chapter is primarily concerned with Edison as an inventor.

6. Hughes, *Networks of Power,* pp. 21–23 *inter alia.*

7. Important contributions in this area, besides Thomas Hughes' *Networks of Power,* include Leslie Hannah, *Electricity Before Nationalisation* (Baltimore: The Johns Hopkins University Press, 1979) and I. C. R. Byatt, *The British Electrical Industry, 1875–1914* (Oxford: Clarendon Press, 1979) on the British experience, and doctoral dissertations (University of Pennsylvania) by students of Hughes, for example, Robert Belfield's on the Niagara Falls systems and Edmund Todd's on German electrification.

11. Hughes, "Edison's Method," p. 10 (see Note 5, Chapter 8).

12. Conot, *Streak of Luck,* p. 91 (see Note 1, Chapter 1).

13. Thomas A. Edison, *The Beginning of the Incandescent Lamp and Lighting System: An Autobiographical Account* (Dearborn, Michigan: Edison Institute, 1976); typescript at the Edison National Historic Site is dated 1926.

Bibliographical Note

1. Edison testimony, *Telephone Interferences,* pp. 10–11.

2. Derganc, "Edison and His Lighting System," p. 54 (see Note 4, Chapter 8).

3. Kruesi testimony, *Edison vs. Siemens vs. Field,* p. 52.

4. N-79-07-12.

5. PN-80-09-23.

6. Memoranda of February 16 and 17, 1881, DF 1881, "Menlo Park Laboratory—General."

7. Josephson, *Edison,* p. 181 (see Note 1, Chapter 1).

8. Frank L. Dyer and T. C. Martin, *Edison: His Life and Inventions,* 2 vols. (New York: Harper and Brothers, 1929): 247.

9. N-80-07-10.

10. Testimony, *Edison vs. Siemens vs. Field,* by Edison, p. 113, and Charles L. Dean, p. 150; and John F. Ott testimony, *Mather vs. Edison vs. Scribner,* pp. 11 and 15.

INDEX

Page numbers in boldface indicate illustrations. When an illustration reference duplicates a text reference, both are given.